Pharmaceutical Nanotechnology

Costas Demetzos

Pharmaceutical Nanotechnology

Fundamentals and Practical Applications

 **Adis**

Costas Demetzos
Faculty of Pharmacy
National & Kapodistrian University of Athens
Zografou, Athens
Greece

ISBN 978-981-10-0790-3   ISBN 978-981-10-0791-0   (eBook)
DOI 10.1007/978-981-10-0791-0

Library of Congress Control Number: 2016938422

Translation from the Greek language edition: *ΦΑΡΜΑΚΕΥΤΙΚΗ ΝΑΝΟΤΕΧΝΟΛΟΓΙΑ. ΒΑΣΙΚΕΣ ΑΡΧΕΣ ΚΑΙ ΠΡΑΚΤΙΚΕΣ ΕΦΑΡΜΟΓΕΣ (Pharmaceutical Nanotechnology. Basic Principles and Practical Applications)* © Parisianou S.A. 2014. All rights Reserved.

The Work was first published in 2014 by Parisianou S.A. with the following title: *ΦΑΡΜΑΚΕΥΤΙΚΗ ΝΑΝΟΤΕΧΝΟΛΟΓΙΑ. ΒΑΣΙΚΕΣ ΑΡΧΕΣ ΚΑΙ ΠΡΑΚΤΙΚΕΣ ΕΦΑΡΜΟΓΕΣ* (Pharmaceutical Nanotechnology. Basic Principles and Practical Applications).

Printed on acid-free paper

This Adis imprint is published by Springer Nature
The registered company is Springer Science+Business Media Singapore Pte Ltd.

This World, The Small, The Great!
Odysseus Elytis
Axion esti – Worthy It Is
(Το Άξιον Εστί, 1959)

*This book is dedicated with respect
and love to the memory of my parents,
Nikos and Maria.
This book is dedicated to those who stood
beside me with a lot of love throughout my
academic career. To my wife and colleague
Voula Dimitriou and to my daughters
Katerina and Maria.*

Foreword

It was in the 1980s that the word nano, used as a prefix to a variety of important-sounding words, reached me from various directions. "What is this nano thing I'm hearing about daily?", I asked a colleague at the university. "I'm worried I'm missing something. I feel as if I'm left behind, in the middle ages of science so to speak". "Don't you know?" he said, expelling streams of tobacco smoke through his nostrils (smoking was not a sin then). "It's the technology of small, dwarfish things. Nano is Greek, it means dwarf". "Thanks for telling me" I said, subtly reminding him I am Greek. I was working on liposomes at that time, small vesicular fatty particles, around 100 nm in diameter, dwarfish in other words. Does this mean I'm doing nanotechnology myself? I wondered. Hope was already enveloping me.

Actually, those of us working on liposomes in drug delivery were not the only ones doing nanotechnology and not being aware of it. Others were doing it as well, for instance those engaged in research with synthetic nanoparticles, dendrimers, viruses and gold particles. As nanotechnology, by convention, deals with systems of sizes in the range of about 1–100 nm, it would also include research on small and large molecules such as sugars, proteins, genetic material and polymers. In fact, work on small, dwarfish structures was so extensive prior to the emergence of the term "nanotechnology", it encompassed a plethora of fields of science and technology. The only thing missing was a name for it. A name that would represent all nano-scale technological activities. And what would that name be? Nanotechnology, obviously! As fate would have it, the first to think of it was Norio Taniguchi, back in 1974.

Nanotechnology is now a huge, ever-growing, ever-expanding discipline. It seems that the name itself has galvanised a massive thrust of related activities. The term nanotechnology has attained such a heavy-weight status in science and technology, it is used by academics and others to name a lab, a department, a new journal, or even to bestow a certain respectability to an activity related to dwarfish things. Nano is king nowadays. Nanotechnology is now taken so seriously, it has predictably attracted the attention of those guarding our health and environment. I refer to research on the toxicology of nanomaterials, usually non-biodegradable ones which, in the form of airborne nanoparticles, were shown to promote fibrosis of the lungs. However, tight regulation by licensing bodies should ensure avoidance of such problems.

Of particular importance to those of us engaged in drug delivery is the role of nanotechnology in advancing the science of therapeutics and their application in the clinic. It has been therefore a great personal pleasure to write the Foreword of the present book by Professor Costas Demetzos. His monograph, *Pharmaceutical Nanotechnology: Fundamentals and Practical Applications*, is a unique publication designed to serve as a scientific textbook for those involved in new technologies, especially nanotechnology of pharmaceutics, as well as to educate students on the nano aspects of the science and technology of pharmaceutics.

This book is divided into three main parts. It begins with an introduction to nano-technology, the physical chemistry, thermodynamics and biophysics of nanosys-tems, offering an analysis of basic facts regarding the size and thermotropic behaviour of systems used in nanotechnology. We learn that nanotechnology deals with the application of technology in the grey area in between classical mechanics and quantum mechanics. It also refers to size-measuring devices that are so minute, they can only be observed under electron microscopy. Remarkably, an existing device of this kind is so sensitive, it is reputed to measure the weight of protons! This is of utmost importance because nanotechnology cannot progress without the means of measuring nano sizes. A section on the historical background discusses the development of modern nanotechnology and the personalities involved. It then deals with the biophysics of a variety of nanosystems, their stability and other prop-erties as well as their physicochemical characterisation.

The author focuses on three different aspects of applied pharmaceutical nano-technology: imaging, diagnostics and therapy. There is extensive discussion on a variety of drug carrier nanosystems, of drug incorporation into such systems, as well as their pharmacokinetics. An all-important consideration of pharmaceutical nanotechnology that enables its application clinically is the evaluation of toxicity and safety of nanosystems. To that end, this book deals with issues of safety, nano-toxicity of nanosystems and the regulatory framework. This includes the procedures that must be followed in order to achieve approval by agencies such as FDA in the USA and EMEA in Europe. Such issues, treated comprehensively in the third part of the book, serve as a reminder of the complexity of steps that must be followed prior to the approval and commercialization of clinically effective drugs.

Pharmaceutical Nanotechnology: Fundamentals and Practical Applications is certain to prove of considerable value to those committed to the science of phar-maceutics and related technologies. It will open new avenues of research and development in pharmaceutical academia and industry. The author is to be con-gratulated for this achievement.

London, UK   Gregory Gregoriadis

Preface

The textbook entitled *Pharmaceutical Nanotechnology. Fundamentals and Practical Applications* is addressed to those in the fields of pharmaceutics, medicine, biology, and life sciences and to those who are interested in materials and new technologies. This textbook includes an introduction to the science of nanotechnology, the basic principles of physical pharmacy, and nanoparticle applications in the fields of diagnostics, imaging, and therapeutics. They are presented in the chapters, the principles of thermodynamics and biophysics as important milestones and as tools for developing and evaluating nanomaterials, nanodevices, and nanosystems. The relationship between the membrane biophysics of artificial nanostructures and the cell membranes' behavior could be studied using thermodynamic parameters' changes, and issues are discussed in terms of the development process of drug delivery nanosystems.

This textbook also presents the applications of innovative nanosystems that are considered as carriers of bioactive molecules, diagnostic and imaging nanotechnology, and relevant nanosystems like liposomes, dendrimers, polymers, polymeric nanocarriers, carbon nanotubes, nanoshells, etc. The toxicity of nanomaterials and nanostructures and the regulatory framework for the approval of the new nanotechnological products are presented at the very last chapters of this book.

Furthermore, I would like to underline in this monograph the usefulness of fundamental sciences such as biophysics and thermodynamics. It is worth mentioning that the investigation of the *mesoscopic* word and of the *metastable phases* of the liquid crystalline state of the matter could be emerged as the building blocks for designing better medicines and improving the health. The principles of basic science can help to rationally design and develop innovative medicinal formulations and to translate the basic scientific principles of manufacturing to improve medicine development process and to meet more efficiently the regulatory requirements. For these reasons, biophysics and thermodynamics could create the science-based platform for *systems therapeutics* and hopefully contribute efficiently to the *cycle of innovation* of nanotechnology-based products, which maps the regulation of new health technology demands.

In other words, my vision is for the scientist, in the field of the development and evaluation of nanosystems that is applied to therapeutics, diagnostics, and imaging, to understand the important position that he holds in the research and development of pharmaceutics, the *systems pharmaceutics*.

The monograph provides fundamental aspects of nanotechnology and fills the gap between nanotechnological systems and functionality of living organisms, providing new aspects on their physicochemical, biophysical, and thermodynamic behavior.

The monograph is addressed to all those involved in recent advances in pharmaceutics. The book is divided into three major parts, each of which is divided into subchapters. Part I refers to the physicochemical and thermodynamic aspects of nanosystems, while their biophysical behavior is correlated with that of the cells of the living organisms. Part II refers to the application of nanotechnology in imaging, diagnostics, and therapeutics. Part III is focused on issues regarding safety, nanotoxicity of nanosystems, and regulatory framework. The monograph promotes the concept that biophysics, thermodynamics, and nanotechnology are considered to be emerging tools that when following the regulatory guidelines provide new and integrated knowledge, for the production of new medicines.

As an academic professor, my vision is that young researchers will use this textbook as a springboard for new knowledge and research for the science of tomorrow. I would like this book to fulfill the verse of Odysseus Elytis, the person who intuitively and unmistakably touched nanotechnology: *"This you should only know: Whatever you save during the lightning, it will remain clean throughout the century"* [(Axion esti – Worthy It Is (*To Άξιον Εστί*, 1959)].

Zografou, Athens, Greece   Costas Demetzos

Acknowledgments

This monograph is the result of a hard and long-term academic effort in the continuously developing field of nanotechnology, in the area of medicines, treatment, diagnostics, and imaging. I have been fortuned enough to have as a mentor, not only a pioneer in the field of nanotechnology, but also a wonderful person, Professor Demetrios Papahadjopoulos, from the University of California, San Francisco (UCSF). He familiarized me with the science of nanotechnology and taught me that the scientific research is a determinant condition for the approach to the truth and the scientific dignity a researcher, a *homo universalis,* must have. I owe my scientific steps of the past 20 years to the acquaintance with the scientist and human, Demetrios Papahadjopoulos. His presence in my life gave me the necessary tools for the scientific being, a refreshing way out to the continuous search of a researcher with social references.

I owe credits to all those who, with their long-term research in our laboratory or with their collaboration, helped us progress in the field of pharmaceutical nanotechnology.

I would like to thank my colleagues Assist. Professor Sofia Hatziantoniou, Assist. Professor Kostas Dimas, and the chemist Aris Georgopoulos for their collaboration.

My special gratitude goes to my colleague Dr. Natassa Pippa for years of her outstanding efforts and fruitful discussions for a long time period to formalize the content of this book and to complete it as a textbook.

I also would like to thank Dr. Aggeliki Siamidi and Dr. Natassa Pippa for their fruitful discussions, the opposite remarks, the uninterrupted research activities, and the continuous research for knowledge, the mutual trust, and their teamwork in order to attribute complex scientific issues.

Finally, I appreciate the fruitful discussions with Mrs. Margarita Parisianou-Papailiou from PARISIANOU S.A., regarding the presentation of this textbook. I would also like to thank Prof. C. Papailiou for his kind consultation on procedural matters.

Contents

Abbreviations

ADC	Antibody drug conjugates
ADME	Absorption distribution metabolism excretion
AFM	Atomic force microscopy
APC	Antigen-presenting cells
ASTM	American Standards for Testing and Materials
ATMP	Advanced therapy medicinal products
ATP	Adenosine triphosphate
AUC	Area under the curve
BBB	Blood brain barrier
BNCT	Boron neuron capture therapy
CAT	Committee for Advanced Therapies
CBER	Center for Biological Evaluations and Research
CDER	Center for Drug Evaluation and Research
C-DOPE	N-citraconyl-dioleoyl-phosphatidylethanolamine
C-DOPS	N-citraconyl-dioleoyl-phosphatidylserine
CDRH	Center for Device and Radiological Health
CHMP	Committee for Medicinal Products for Human Use
CMV	Cytomegalovirus
CNT	Carbon nanotube
COMP	Committee for Orphan Medicinal Products
CPD	Continuing Professional Development
Cryo-AFM	Cryogenic atomic force microscopy
Cryo-TEM	Cryogenic TEM
CTA	Center for Technology Assessment
CV	Coefficient variation
CVM	Center for Veterinary Medicine
CVMP	Committee for Medicinal Products for Veterinary Use
DAPE	Diacetylenic phosphatidylethanolamine
DDS	Drug delivery system
DLS	Dynamic light scattering
DMA	Dynamic mechanical analysis

DOPC	Dioleoyl phosphatidylcholine
DOPE	Dioleoyl phosphatidylethanolamine
DPPC	Dipalmitoyl phosphatidylcholine
DPPG	Dipalmitoyl phosphatidylglycerol
DSC	Differential scanning calorimetry
DSPG	Distearoyl phosphatidylglycerol
DTA	Differential thermal analysis
DTG	Differential thermogravimetry
DTPA	Diethylene triamine pentaacetic acid
EGA	Evolved gas analysis
ELISA	Enzyme-linked immunosorbent assay
EM	Electron microscopy
EMA	European Medicinal Agency
EPC	Egg phosphatidylcholine
EPR	Enhanced permeability and protection
FDA	Food and Drug Administration
FFEM	Freeze fracture electron microscopy
GMP	Good manufacturing practice
GTMPs	Gene therapy medicinal products
HAI	Health Association Infection
HMPC	Committee on Herbal Medicinal Products
HMV	Heating method vesicles
HPE	Handbook of Pharmaceutical Excipients
HSPC	Hydrogenated soy phosphatidylcholine
ICH	International Committee for Harmonization
ICTAC	International Confederation for Thermal Analysis and Calorimetry
IMI2	Innovative Medicines Initiative 2
LFM	Lateral force microscopy
LLD	Liposomal lock in dendrimers
LTD	Ligand-targeted liposome
LUV	Large unilamellar vesicles
MAAs	Marketing Authorization Applications
Mab	Monoclonal antibody
MEMS	Microelectromechanical system
MLCRS	Modulatory Liposomal Controlled Release System
MLV	Multilamellar vesicle
MPS	Mononuclear phagocyte system
MRAM	Magnetoresistive random access memory
MRFM	Magnetic resonance force microscopy
MRI	Magnetic resonance imaging
mtDNA	Mitochondrial DNA
MVV	Multivesicular vesicles
MWNT	Multi-wall nanotube
NAS	National Academy of Science
NASA	National Aeronautics and Space Administration

NASBA	Nucleic acid sequence-based amplification
NCI	National Cancer Institute
NCL	Nanostructured lipid carriers
NCST	National Strategy for Combating Terrorism
NFAS	Non-flame atomic spectroscopy
NIH	National Institute for Health
NMR	Nuclear magnetic resonance
NNI	National Nanotechnology Initiative
NSOM	Near-field scanning optical microscopy
O/W	Oil in water
OLV	Oligolamellar vesicles
ORA	Office of Regulatory Affairs
PCR	Polymerase chain reaction
PDCO	Pediatric Committee
PDI	Polydispersity index
PE	Phospahtidylethanolamine
POPE	Palmitoyl-oleoyl-phosphatidylethanolamine
QbD	Quality by design
QD	Quantum dots
RAM	Random access memory
RES	Reticuloendothelial system
REV	Reverse phase evaporation
RIA	Radioimmunoassay
RSV	Respiratory syncytial virus
RT-PCR	Real-time PCR
SARS	Severe acute respiratory syndrome
SEM	Scanning electron microscopy
SFM	Scanning force microscopy
SLN	Solid lipid nanoparticle
SPARC	Secreted protein acidic and rich in cysteine
SPM	Scanning probe microscopy
STM	Scanning tunneling microscope
SUV	Small unilamellar vesicle
TEM	Tunneling electron microscopy
TEPs	Tissue-engineered products
TG	Thermogravimetry
TMA	Thermomechanical analysis
W/O	Water in oil
W/O/W	Water in oil in water
XPS	X-ray photoelectron spectroscopy

List of Figures

List of Tables

Part I
Physical Pharmacy: Biophysics and Thermodynamics

Chapter 1
Introduction to Nanotechnology

Abstract Nanotechnology is a multidisciplinary scientific field that deals with the development and use of materials that can be used to produce devices and products with dimension equal to one billionth of a meter. One nanometer (nm) equals one billionth of a meter ($10^{-9} = 0.000000001$). Nanotechnology refers to the science and the technology where the structural units of the matter used in the formation of complex macromolecular systems are in nanoscale. Pharmaceutical nanotechnology is the application of nanotechnology to pharmaceutics in life and health sciences and is based on nanostructured biomaterials, which promote innovative drug delivery systems for therapeutic purposes, advanced diagnostic biosensors, and imagine agents that are dealing with the early diagnosis of the diseases. Nanotechnology is also applied in physical engineering, in electronics, and in the environment and technology.

Keywords Nanotechnology • Nanometer • Nanometrology • Richard Feynman • Nanostructured biomaterials

1.1 Introduction

Nanotechnology is a promising multidisciplinary scientific field that deals with the development and use of materials that can be used to produce devices and products with dimension equal to one billionth of a meter. It has evolved in a wide variety of nanosystems [2]. The nanoparticles and the nanomaterials have new properties that depend on their dimensions. In science and technology, the prefix *nano-* comes from the Greek word νάνος=*nanos* that means something very small. One nanometer (nm) equals one billionth of a meter ($10^{-9} = 0.000000001$). Nanotechnology refers to the science and the technology where the structural units of the matter used in the formation of complex macromolecular systems are in nanoscale. An interesting approach in defining nanotechnology is reported in M. Saladin El Naschie's [7] article that was published in the scientific journal *Chaos, Solitons and Fractals* in 2006. According to the author, the term nanotechnology corresponds to the processes in *physics, chemistry, and biology in dimensions of one billionth of meter and defines nanotechnology as the application of the technology in the gray area between*

© Springer Science+Business Media Singapore 2016
C. Demetzos, *Pharmaceutical Nanotechnology*,
DOI 10.1007/978-981-10-0791-0_1

the classical mechanism and quantum mechanics. This *gray area*, according to the opinion of theoretical physicists, is defined as the *mesoscopic* area, and the systems that are developed in this area are called *mesoscopic systems.* According to this approach, theoretical physicists define what we call nanoworld. The definition of nanotechnology which is published in *Towards a European Strategy for Nanotechnology* [Brussels, 12.5.2004 (COM) 338 final] from the European Commission is concise and covers all the fields of nanoscience. Nanoscience refers to the study, the measurement, and the understanding of the interactions of materials in nanoscale. Physics, chemistry, biology, mathematics, the science of materials, pharmaceutics, and engineering take part in the scientific and research field of nanosciences that can be characterized as the interdisciplinary approach of the nanotechnology applications. According to the National Nanotechnology Initiative (NNI) strategic plan in February 2014 (http://www.nano.gov/node/1113), *nanotechnology is the understanding and control of matter at dimensions of roughly 1–100 nm, where unique phenomena enable novel applications.* The nanodimension is not the smallest. Smaller dimensions than that of nanometer are the dimension of angstrom (0.1 nm) and the prefixes *pico-, femto-, atto-,* and *zepto-* that relate to the 1/1,000 mathematical degradation of dimensions of matter from the nanodimension. Regarding the safety of nanotechnological products, there are objections concerning the risk effects of the human health from the use of nanoproducts (see Chap. 6) (*European Commission, Nanoscience and Nanotechnologies: An Action Plan for Europe, 2005–2009 COM, European Commission, Brussels 2005. http://cordis.europa.eu/nanotechnology/nanomedicine. htm*). The European Union agenda (EU 2020) are well fitted with the challenges and opportunities that nanotechnology provides, for improving societal challenges, to make daily human activities and life better, and to manage the economic growth in a more rational manner (*European Commission, Directorate-General for research and Innovation Industrial Technologies. Nanosciences, Nanotechnologies, Materials and new Production Technologies (NMP) 2013, EUR 13325 EN*).

1.2 Brief Historical Overview

The term nanotechnology is very broad and general and includes anything within the dimensions of a nanometer. Therefore, it can be divided into more specialized areas like nanoelectronics, nanomaterials, medical diagnostic tools and sensors, flexible display technologies and e-papers, composites containing nanotubes, printable electronic circuits, etc. [2]. The applications are countless, while the effects are perceptible in many levels, especially in the economy affecting industries and economies, but also in the social sector improving significantly our living standards. The origin and development of nanotechnology are unclear. The first nanotechnologists might have been the glass technicians in medieval times that used furnaces to mill glass. The glass technicians could not understand how processing gold created colors. The process of engineering nanosystems and nanostructures, especially in the production of gold quantum dots, was used in the Victorian and medieval churches that were famous for the lovely stained glass windows and in the ancient Ottoman mosques that were famous for the enamel surfaces. The different color variations and materials' behaviors that are being used in nanosystems are based on the fact that the features of materials in nanoscale are different from those in microscale dimension. The achievements of sciences (physics, biology, chemistry) but also the evolution in technology and the progress in the science of materials have helped us gather information to understand nanotechnology. These developments turned out to be a necessity for the orientation, the selection, and the efficacy of these applications in this particular field.

The first scientific report of nanotechnology (without the use of the name "nanotechnology") was made during the speech of Richard Feynman entitled *There's plenty of room at the bottom* during the American Physical Society dinner in 1959 [5]. In his speech, he announced nanotechnology, many years before the use of the prefix *nano-*. Also, he discriminated the methods of constructions of nanostructures:

• Top-down technology
• Bottom-up technology

Top-down technologies are the ones applied to industries these days. On the other hand, bottom-up technologies form systems from atoms or molecules. The most important challenge of this method is the ability of formation and evaluation of structures, the way nature creates them in dimensions of nanometers [8]. It is obvious that the physical phenomena in these dimensions are the one that lead the research and development, like nature requires. The evaluation of these new materials and their application to everyday life, and especially in the field of health, render the technology of nanodimensions as a new field of research and applications.

Richard Feyman's speech triggered a discussion which was based on the idea that the whole of *Encyclopedia Britannica* could be written on the top of a pin. He also mentioned the accurate copy of certain individuals, the size shrinkage of the computers (computers were much bigger than today, but possibly he meant even smaller), and a way of developing atom and molecule handling ability directly, having tools in one tenth of the size, exactly like the ones we have in any engineer workshop. These small tools would help a new generation of toolkits of one centimeter being developed and used. As the size would decrease, the tools would have to be redesigned as the relative resistance of some forces would change. The gravity would have less meaning but the surface tension and the *van der Waals* forces would be more important. Feynman considered only one obstacle for the management of the matter in molecular and atomic level: the lack of experimental instruments that could have been used for nano-level measurements. Actually, the term nanotechnology was first introduced by Professor Norio Taniguchi from the University of Sciences in Tokyo, in 1974, in his thesis entitled *On the basic concept of Nano- Technology*, where he precisely described the formation of materials with dimensions of a nanometer. During the 1980s, the term appeared again and was redefined by Eric Drexler, the president of Foresight Institute in his book *Engines of Creation: The Coming Era of Nanotechnology* published in 1986, making an effort to secularize the term nanotechnology. This way of popularizing was connected with his effort to describe tiny machines, of nanometer scale, that would have the ability of self-assembly and would function in all levels of human activities. He also looked deeper into this matter in his doctoral thesis, *Nanosystems*: *Molecular Machinery Manufacturing and Computation* emphasizing to all technical issues [4].

Eric Drexler is well known as the scientist who brought the revolution of nanotechnology to today's point. He increased world's recognition related to science, trained those who would research and develop nanotechnology in the future, and lightened up this field. For these reasons, he was granted the first PhD in nanotechnology. In 1992, Drexler urged a legislature committee to take this matter under consideration. He also wrote plenty of books related to nanotechnology. *Nanosystems*: *Molecular Machinery Manufacturing and Computation* was the first one written in 1981. Another book that was published later on, in 1986, was the *Engines of Creation*: *The Coming Era of Nanotechnology*. Another book *Unbounding the Future*: *The revolution of Nanotechnology* was written in 1991.

The calculation methods play an important role in the field of nanotechnology nowadays because nanotechnologists can use them to design and simulate a wide range of molecular systems. In various conversations on nanotechnology, the creation of a computing algorithm with a wide range of abilities for the construction of various molecular structures has been mentioned. The ability of self-similarity was related with the idea that the computing programs could recreate more programs by the principle of self-similarity. As a result, nanotechnology would affect the economy and the prices of products whose production was related to computers and automated processes depended on computing programs. Smaller structural elements of a product would have been manufactured in assembled series that would have been assembled gradually in bigger series until the completion of the final product. This logical way of thinking could create new business data and would affect production, quality, speed, and obviously the cost of products, while a new regulatory framework for protection would be a necessity. The chronological evolution of nanotechnology through historical milestones is present in Table 1.1.

1.3   Nanometrology

To calculate the mass of nanosystems, researchers from the Catalan Institute of Nanotechnology in Barcelona, Spain [3], have created the smallest and most accurate nanodevice of the world that is so sensitive that could also weight the mass of protons (the mass of a proton is 1.7 yg; 1 yg = 10^{-24} g). The sample is placed in the central of a carbon nanotube with the length of 150 nm. The nanodevice is called nanomechanical mass sensor. Similar nanodevices have been used in the past, to measure the mass of isolated cells, viruses, or even isolated molecules. This carbon nanotube acts like a guitar string that oscillates in a very high frequency of 2 GHz (two billion cycles per second). The objects to be weighted are placed on the carbon nanotube, changing its eigenfrequency and forcing the nanotube to vibrate in a different frequency. The sensor can accurately calculate the mass of the sample by

Table 1.1 Important dates in the evolution of nanotechnology and of nanotechnological products and medicines

1905	Einstein publishes his study on the dimension of sugar molecule, approximately 1 nm
1959	R. Feynman, in his lecture at the annual meeting of American Association of Physical Sciences, claims that *There is Plenty of Room at the Bottom*
1974	Norio Taniguchi introduced the term nanotechnology
1981	E. Drexler designs molecular machines that mimic enzymes and ribosomes
1984	The first description of the term "dendrimer" by D.A. Tomalia and the preparation method of PAMAM dendrimers
1991	The discovery of carbon nanotubes (CNTs)
1994	Drug delivery systems
1995	FDA approved Doxil® (liposomal doxorubicin)
1997	FDA approved AmBisome® (liposomal amphotericin B)
1998	DNA nanoparticles for controlled gene delivery
2000	The first FDA approval of medicinal product based on the technology of Liquid Crystals (NanoCrystal® Technology) and the solid dose formulation of the immunosuppressant sirolimus – Rapamune®
2005	FDA approves Abraxane®, the nanotechnological formulation of paclitaxel
2008	In market: PEG-Certolizumab pegol (trade name, Cimzia®) anti-TNF Fab for rheumatoid arthritis and Crohn's disease
2012	Biomimetic drug delivery systems: the first publications in literature
July 2015	Successful clinical trials of ThermoDox® (*lyso-thermosensitive liposomal doxorubicin*)
October 2015	FDA approves Onivyde® (irinotecan liposomal) for advanced pancreatic cancer

measuring the frequency change. To obtain accurate results, the sensor must function under controlled conditions: the nanotube must be totally clean at $-289\ ^{\circ}C$, while the sample must be placed in vacuum, free from mechanical and electrical changes. The researchers from the Catalan Institute of Nanotechnology used this tiny nanodevice that they had built, not only to weight very small charges but to study the interaction of these charges with the carbon nanotube as well. For example, they measured the rate of the naphthalene absorption on the surface of the nanotube. In the future, researchers believe that sensors of the same level would improve the scientific fields of mass spectrometry, magnetometry, nanometrology, and surface sciences. Also, the sensor could have been even more sensitive, if a very small sample reception could be manufactured, so the sample would be placed in a particular spot on the nanotube, restricting frequency fluctuations.

1.4 Nanotechnology in Physical Sciences and Engineering

In the future, the effects of nanotechnology will expand into plenty of aspects in everyday life, and its importance is based in the following fields [announcement of the committee of European Commission *"Towards a European Strategy for*

Nanotechnology" (Brussels, 12.5.2004, COM (2004) 338 final, http://cordis.europa. eu/nanotechnology), in the field of medicinal applications of therapeutics and diagnostics of diseases. Bio-inspired systems transferring bioactive molecules with sensors aiming the diseased tissue are the evolution on the field of therapeutics (see Chap. 5). The application of inorganic nanoparticles with the method of thermal healing is also a field in cancer therapy. The technological achievements could improve implants with nanomaterial coating. Biomimetic materials with self-assembly abilities and self-similarity properties can lead to the development of artificial organs. The contribution of nanotechnology in life sciences and health will be thoroughly mentioned in the following chapters of this book [8]. In the field of information, its ability of storing information (e.g., 1 terabit/inch^2) and the prospect of biomolecular nanoelectronics and quantum technology of electronic computers can lead to challenges and launch new projects in information transport. Energy storing and saving are important application fields of nanotechnology. The storage of hydrogen as a combustible fuel can be accomplished by solid nanostructures of small molecular weight and high dynamic energy storage. In the science of materials, there are a lot of nanotechnological applications. The surface nanomaterial modification cannot be scratched and consequently become waterproof, cleaner, and sterile are just a few of the applications in the science of materials. The production of nanosensors and molecular electronic devices are new application field of the new materials' abilities in nanometer scale. In the field of economics, nanotechnology will aid in the production of new materials and biomaterials for medicinal purposes with lower cost due to the replacement of the conventional technologies with the new ones. For example, a television of a cathode tube consumes much more energy than a liquid crystal or plasma television. Similarly, the devices that will incorporate nanotechnology will be functioning with less cost and energy consumption. Solar energy, from a non-exploitable/experimental source of energy, will become sustainable and profitable and will be used even in aircrafts and ships. In the field of development and discovery of new technologies, carbon nanotubes are the first-class material for the creation of a compact construction. Gradually, the majority of the building materials will be based on carbon nanotubes, and this will create an opportunity for the construction of very high buildings, bridges, etc. Nowadays, National Aeronautics and Space Administration (NASA) experiments on a new way of uplifting various satellites and space rockets to space. This will be accomplished through special ropes and will become a new type of elevator. In the field of clothing, everyday clothes will not wrinkle or get worn out. The spectacular invasion of nanotechnology in clothing took place in 2002, when Levi Strauss & Co. used the stain repellent Teflon, of the chemical industry DuPont. When the stain contacts the cloth, millions of nanofibers repel it. The nanofibers are incorporated onto the fabric when it is immersed into the special chemical solution. This success led to the use of this technology, and nowadays there have been clothes made from nanofibers that function as sweat pumps, absorbing body moisture and leading it to the surface of the fiber until it is dries out. In the field of everyday hygiene, simple sprays will pervade the air with oxygen or with flower scent, eliminating bacteria. This can happen automatically through air conditioning systems for a world free from unpleasant scents. Similarly, toilettes will not be places of contamination and infections. Toothbrushes with nanoparticles will take care of dental hygiene. Night creams will restore cells to their original condition.

1.5 Nanotechnology in Life and Health Sciences

The nanotechnological applications and the research activities that exhibit great interest are on the medicine development field, therapeutics, diagnostics, and imaging. Serious metabolic diseases, like diabetes, and others, like cancer, could be taken under control with the use of nanotechnology [6]. Nanotechnology will offer tools for genetic tests, and the administration of innovative medicines will contribute to the improvement of diagnostics and therapy. The nanostructures and nanosystems that are able to incorporate bioactive molecules will be an innovative approach for biomolecules' delivery and targeting (see Chap. 5). Also, the alliance between nanotechnology and the systematic approach of biosciences and pharmacology, through the -**omics** (genomics, proteomics, metabolomics), can lead to the development of nanostructures and, therefore, nanodevices that will function according to each patient genetic print. This way, we are led to the development of new medicines and personalized therapeutic and diagnostic approach to diseases. Also, in a macroscopic level, we can highlight that the above approaches can lead to the development of pharmaceutical products that will contribute to better public health (see Chap. 3).

In 1990, IBM placed thoroughly and carefully 35 xenon (Xe) atoms for the three-lettered syllable of the business that was the smallest company logo in the world. At that point, scientists at Cornell University produced a visible "nanoguitar" that could not be seen with bare eyes. The strings, consisted of a few atoms, could be played with laser beams to produce 17 octaves more from the typical guitar, something that was above the human hearing ability. After the creation of the nanoguitar, in 1999, completed chips of 1 molecule width were developed. These were the smallest ever made, produced from the scientists of Hewlett-Packard Company in California. The development of that completed chips was announced in *Science* magazine, in July 16, 1999. In slightly bigger dimensions, there are enough fields for development in the near future in nanoengineering artificial organs. We already know that we can connect appliances with parts of our nervous system that process optic or hearing information. Where we fail is usually at the scale part – we cannot construct devices that can see or hear and be so small, like the ones that nature creates. Nanotechnology can turn these data around. But the technological method that will present the first results is most likely going to be painfully slow and, therefore, expensive. However, the possibility of replacing senses like vision, hearing, and touch definitely exists. The implantation of these appliances to our nervous system though still retains a challenge as we need to avoid classical problems of implantation rejection. Even if we could create a more capable eye, for example, how could we connect it to an adjustable brain that could receive (and process) the information of a normal eye? As an example that correlates the application fields that we are concerned for, materials with improved durability and synthetic surface properties offer the possibility for implant use in every way, from artificially hearts to hips.

Nanotechnology enters dynamically into the battle against cancer. The American Cancer Institute announced a program for the development of the usage of tiny "tools" targeting cancer. According to specialists, the development and use of devices of molecular size offer us new ways for tracking, diagnosing, and treating cancer in the early stages and without any side effects. Nanotechnology is already being used in medicine, mostly with "devices" in molecular size like antibodies. The engagement of nanotechnology in the biological material level (proteins, antibodies, biological macromolecules) regarding the science of biotechnology can be attributed with the term nanobiotechnology. Molecular biology through nanotechnology can create a whole new field of research in the health sector with the use of basic physical sciences that are enclosed into the term nanobiotechnology. Also, we can mention that in the field of research of new medicines, the nanobiopharmaceutics mark the opening of new horizons in the discovery of new therapeutic approaches and new diagnostic tools. Nowadays, the prefix *nano-* is extremely important and the majority of scientists turn their research activity into the field of nanotechnology. The nanomaterials can be devices or systems, supramolecular structures, and complex or composite materials.

The pioneer of nanotechnology, Albert Franks, defines nanotechnology as the *scientific and technological field that deals with size and tolerance of 0.1–100 nm.* In nanoscale level, the materials' physical, chemical, and biological properties have fundamental differences and present unexpected properties in relation to the initial material, because the quantum mechanics interactions in the atomic level are affected by the size of the material in nanoscale. The nanodevices have the same size as the biological entities. Nanodevices with size less than 100 nm are smaller than that the human cell and its organelles and can be compared size wise with biomolecules like enzymes and receptors. For example, hemoglobin has a diameter of about 5 nm, while the cell lipid membrane has width of about 6 nm. Nanodevices of less 50 nm can easily enter most cells, while those less than 20 nm can easily enter blood vessels, making nanodevices capable of penetrating biological barriers.

Nanosciences and nanotechnology compromise new approaches to the research field. These fields have the ability of understanding new phenomena and show new properties of these materials. The real challenge is that the nanostructures that are created do not conform to the classic laws of physics due to their size, but are too big to follow the principles of quantum mechanics. These systems belong to a scale that is called *mesoscale*, as we mentioned at the beginning of the chapter. As a result, there is the appearance of a new scientific field, despite of the scientific approaches of the past centuries in the dimensions of *macro-* and *micro-*world. This way the science of *mesoscale* and *mesoworld* is the new challenge, where the *nano*dimension is the ruling size for the creation of structures and systems with new properties. The interfacial phenomena in this dimension are the ones ruling due to the large overall surface of the nanosystem and due to the fact that most of the atoms of these materials are on their surface.

These approaches can result to the development of new pharmaceutical and diagnostic products. The research for development of new medicines has two directions. The first one deals with the identification of new bioactive molecules and the second deals with their safety and effectiveness. It is obvious that the technological platforms that will be used, as well as the development process for producing new medicines, should meet the requirement criteria of the regulatory agencies.

To a large extent, research has turned toward the improvement of the already known bioactive molecule through the use of new pharmacotechnological formulations, without abandoning the approaches mentioned earlier. The developments of the last decade in the science of materials and biology (especially molecular biology) had led toward the use of new pharmacotechnological formulations that directly affect the health sector and the involved institute bodies (organizations for medicinal and health product approval), pharmaceutical industry, research, and scientists.

Nanotechnology correlates directly with health sector, sciences like pharmaceutics, chemistry, mathematics, physics, biology, and pharmacology, and is applied to patients through clinical practice. Nanotechnology in the field of health is based to three overlapping technologies of molecular sciences. Nanobiomaterials can contribute to the development of diagnostic biosensors, in therapeutics, and in producing smart nanodevices that are able to behave as nanorobotic systems. Molecular biology and pharmacology through the developments in genomics, proteomics, and metabolomics can be correlated with pharmaceutical nanotechnology and lead to personalized medicine or the population approach of therapy.

Nanotechnology promises to treat incurable diseases and make medicine easier. Nanorobots, too small to be seen by the human eye, will enter the human organism and give advanced diagnosis for the condition of the organism. Then other nanorobots will manage the treatment. A small blood donation from the fingertip will be sufficient for full blood tests. The treatment will be targeted only to the diseased cells and will not cause any side effects. The creation of nanostructures in molecular level aiming the self-assembly that would function in molecular level but also in tissue regeneration is a prerequisite for nanotechnology in the health sciences. Nanotechnology and its application in pharmaceutics is expected to offer important advantages to biomedical applications like drug administration, diagnostics and imaging, gene therapy, and cancer therapy.

Here are the main directions that pharmaceutical nanotechnology offers to various sections:

- Nanotechnology in imaging and diagnostics has dealt with the development of new diagnostic nanodevices that require small amount of sample and give accurate results in short term.
- Multifunctional medical devices with a decrease size that are able to enter into the human organism to ensure painless use and achieve biocompatibility with the human organism.
- Controlled release nanodevices that are able to deliver bioactive molecules to the target tissues. The main concern of nanomedicinal products regarding the

controlled release process of the bioactive molecules is to improve their bio-availability and effectiveness and their pharmacokinetic and pharmacodynamic profile and to reduce possible toxicity and immunogenicity.

At this point, the extensive use of nanotechnology in the science of cosmetology should be mentioned. The industry of cosmetics is one of the most powerful in the field of nanotechnology. L'Oreal has dozens of patents for cosmetic products that use nanotechnology. Because of these patents, this company is rated among the first ten companies that have patents in the field of nanotechnology, outraging some of the biggest companies like General Electric and Motorola. Products like emulsions, creams, sunscreen products, makeup products, etc. follow the development and the tests according to the behavior and the physicochemical properties of nanoparticles. Not only L'Oreal but other big companies like Estee Lauder, Christian Dior, and Shiseido invest in the field of cosmetics using nanotechnology. From all of the above, it is clear that the benefits from the development of nanotechnology in the field of pharmaceutics are very important, emphasizing in the development of nano-technological systems for the control release of bioactive molecules (Chap. 5), having all the advantages that were previously mentioned.

1.6   Nanotechnology in Electronics

The value of small dimensions is very important. In the end of the 1940s, the first computers were invented but they were extremely complicated, intractable, and huge. They were too slow and they consumed great amounts of energy. In 1947, J. Bardeen and W. Brattai who worked in Bell Telephone Laboratories invented the transistor, and later on, W. Shockley managed to improve it. Transistors could be used either as switches or as amplifiers in a circuit. In 1956 he received the Nobel Prize for this breakthrough innovation that led the way toward the shrinkage of mate-rials. In 1965 Gordon Moore, cofounder of Intel Corporation (a company that made Pentium microprocessors), suggested two predictions for the computer processing strength. The first suggestion, known as the "Moore's law," predicted that the num-ber of transistors incorporated into a processor would double up every 12 months. As it turned out, this prediction was not true. Moore had to reconsider many times the time period, and in 1990, he ended up suggesting that the number of transistors incorporated in a complete chip – and not in a processor – would double up every 18 months. "Moore's law" even though it was inaccurate plays a very important role, as it describes the tendency of the past decades in the field of computers. The first Intel processor (1971) had just 2,250 transistors, while a modern computer has doz-ens of millions and those in a few years' time will have a billion transistors.

Moore's law in general was about the upward trend, not only in transistors but in random access memory (RAM), in hard drives, etc., where the abilities kept on growing. The second law was talking about shrinkage of completed circuits, pre-dicting that they will get smaller and smaller as time passed by. This prediction was confirmed from the developments that counter the first one. The size of the complete circuits and transistors gets four times smaller every 3 years. If this rate is continuous, which is very likely, in a few years' time, the circuits will be some millimeters wide.

The shrinkage and the upgrowth of transistors could not be continued forever because of physics, functionality, production, finances, etc. that put a limit to Moore's law. A very good example is silicon, the natural material that transistors are made of. Silicon has natural limits (strength, shrinkage, good functionality, etc.) that cannot be broken. All the other electronic technologies that computers are using are at the same state. Everything comes down to the fact that electronic technology has an expiration date. This means that the developments in the field of electronics will not be continued, unless something else is found.

At this point, nanotechnology is a possible solution to overcome these problems, and *Moore's laws* will be valid in the form of perpetual development. From all the up-to-date scientific researches and lab work, the most important nanotechnological changes in informatics will be the following:

- Carbon nanostructures (graphene) will replace the silica in transistors, while each transistor will have one and only electron.
- The magnetoresistive random access memory (MRAM) will be developed and will be able to store and preserve all the data magnetically.
- The new technology of spintronics will be developed and eventually will replace the electronics.

1.7 Nanotechnology in the Environment and Technology

The contribution of nanotechnology in water resources and environment can be of value with the development of scanning tools or elimination of harmful microorganisms and pesticide decomposers. The development of photocatalytic techniques based on nanotechnology could help eliminate environmental contamination and pollution. It has to be mentioned that the study of the property of the matter in nanoscale is affected from the evolutions of scientific instrumentation. An example is the invention of scanning tunneling microscope (STM) that contributed to the development of nanotechnology. Nanotechnology is based purely in nature, but the abilities of living nature are limited: it cannot handle high-temperature conditions that are required for ceramics or metal pipes. On the other hand, modern technology has in possession a wide range of artificial conditions – excellent material purity, extreme cold, vacuum conditions – where matter reveals amazing properties. Quantum phenomena are such properties that in most cases are in contrast to what we know in our everyday life. This way the particles of the nanoworld have wave properties at the same time. For example, it is possible for an atom to go through between two empty spaces at the same time, just like a wave, and then emerge at the other side.

The particles have complete new abilities, as their size reaches the nanometer: the metals become semiconductors. Complete undetected substances like cadmium telluride (CdTe) fluorescent with all the rainbow colors, while other elements convert light to electric power. The percentage of atoms in elements' outer surface increases when the elements are in nanosize. The atoms in the outer surface have different abilities than the ones in the center of the element and are more prone to counteractions. For example, gold, in nanometric scale, is a very good catalyst for fuel cells. Also, nanoparticles can be coated with other substances.

According to the European Committee, there are often reports on the revolutionary potential of nanotechnology, i.e., to the possibility of the impacts in industrial production methods. Smaller, lighter, faster, and more efficient materials, constructive elements, and systems that nanotechnology has to offer can give solutions to many current problems. This way, new opportunities for wealth and employment are appearing.

Moreover, nanotechnology is expected to significantly contribute to the confrontation of global and environmental challenges, as product implementation and readjusted processes in particular uses, resources, and waste saving, as well as pollutant emission reduction, will be possible. At the time being, there is a huge global progress of nanotechnology due to competitiveness. By the end of the 1990s, Europe had invested on time on many nanoscience programs and developed a solid base of knowledge. Now, it is time for the European industry and society to take advantage of this knowledge and develop new products and processes. Nanotechnology was the basic topic of European committee toward a strategy to nanotechnology announcement. In this announcement, it was suggested not only to reinforce nanoscience and nanotechnology research but also to take into consideration other interdepended dynamics.

- National research programs and investments to ensure that Europe has teams and infrastructures that will be able to compete in international level should be established. The public and private sectors in Europe should collaborate in order to achieve a sufficient scientific mass.
- Competitiveness factors, like metrology, the settings, and the copyright that will lead to industrial innovation and competitive advantages for large, average, and small businesses, should not be omitted.
- Continuing Professional Development (CPD) [1] are considered as very important factors. Europe could be able to provide room for scientific business and a positive attitude of the production engineers toward these changes. A purely interdisciplinary research in nanotechnology, demands new approaches to education and training for research.
- Public information and communication and the risk assessment in health and in environment are key points to ensure that the development of nanotechnology meets public expectations. The majority of scientists (pharmacists, chemists, physicists, material scientists, etc.) orient their research and applications toward nanotechnology.

1.8 Summary

Nanotechnology deals with materials that produce systems with dimension equal to one billionth of a meter.

The first report of nanotechnology was from Richard Feynman but Professor Nario Taniguchi was the first to introduce the term nanotechnology.

The fields of applications of nanotechnology are physical sciences and engineering, life and health sciences, as well as electronics.

References

1. Brooksbank C, Janko C, Johnson C et al (2015) LifeTrain: driving lifelong learning for biomedical professionals. J Med Dev Sci 1(2):1–7
2. Cauvreur P, Vaultier C (2006) Nanotechnology: intelligent design to treat complex diseases. Pharm Res 23:1417–1450
3. Chaste J, Eischler A, Moser J et al (2012) A nanomechanical mass sensor with yoctogram resolution. Nat Nanotechnol 7:301–304
4. Drexler E (1992) Nanosystems: molecular machinery manufacturing and computation. Wiley, Chichester
5. Kewal KJ (2008) The handbook of nanomedicine. Humana Press, Basel
6. Mauro F (2005) Cancer nanotechnology: opportunities and challenges. Nat Rev Cancer 5:161–171
7. Saladin El Naschie M (2006) Nanotechnology for the development word. Chaos Solitons Fractals 30(4):769–773
8. Whitesides GM, Grzybowski B (2002) Self assembly at all scales. Science 295:2418–2421

Chapter 2
Fundamentals of Physical Pharmacy: The Biophysics of Nanosystems

Abstract The principles of physical sciences and the application of physical laws in the field of medicines and in pharmaceutical sciences are considered as effective tools in order to develop nanotechnological products. Biomolecules and biomaterials that are the building blocks of nanosystems and their similarity with the structural elements of the human cell are important factors for the understanding of the physicochemical behavior of nanosystems. Biophysics and thermodynamics of cell membranes reflect to the behavior of nanoparticulate systems that are able to deliver bioactive molecules to the target tissues. The liquid crystalline state of nanosystems leads their behavior, while their thermotropic properties can provide information regarding their physicochemical profile and consequently their therapeutic effectiveness. Their stability is considered as a crucial issue, and DLVO theory efficiently explains their behavior and provides evidence that corresponds to their behavior in in vitro media and in vivo experiments. Thermal analysis is also a useful technique that is used to measure thermodynamic parameter that project to their thermotropic behavior. Freeze-drying process is an extensive studied technique that applied in dispersed systems to secure their lifelong physicochemical stability.

Keywords Physical pharmacy • Biophysics • Thermal analysis • Microscopy • DLVO theory • Stability

2.1 Fundamentals of Biophysics in Liquid Crystalline State of Nanosystems

2.1.1 Introduction

The application of physical laws related to the design and development of effective nanotechnological products in the field of medicines demands knowledge of the basic principles of physical sciences. The physical parameters that relate to the behavior of nanodevices but also to their thermodynamic evolution during time are a necessary tool and a prerequisite for their evaluation. The science of biophysics is also an important approach for understanding the physicochemical behavior of nanosystems that are used in therapeutics, diagnostics, and imaging of the diseased

© Springer Science+Business Media Singapore 2016   17
C. Demetzos, *Pharmaceutical Nanotechnology*,
DOI 10.1007/978-981-10-0791-0_2

tissues. The structural unit of human cell through the logic of biomolecules and biomaterials completes the chemical imprint of nanosystems. The evaluation of physicochemical characteristics of biomaterials that are chosen to be biocompatible and biodegradable for nanosystem design is evolving continuously through new scientific and technological approaches.

Science and technology produce new tools, like electronic microscopy for the study of the nanoparticulate shape, light scattering techniques for the study of their physical characteristics, and spectroscopic techniques for the study of interaction between the biomaterials that are composed of and for other parameters. The study of the thermodynamic parameters of nanosystems for the understanding of their thermotropic behavior in a biological environment is also important. The thermal analysis techniques that are going to be mentioned in this chapter are necessary tools for the study and evaluation of phase transitions, of the liquid crystalline state. The data collected are important because, in a biological environment, the viability and the nonlinearity of nanosystems demand multifunctional approaches. The liquid crystalline state that we are also going to mention is an autonomous state of matter and is considered to be extremely important as a biological, thermodynamic, and physicochemical entity. The study of the transition of the liquid crystalline state of nanosystems simulates the one in biological systems and contributes in their design and development, aiming toward the bioactive molecule transport to the targeted tissues [1].

Physical pharmacy [1] is the application of physical sciences principles and laws in the study and evaluation of medicinal products. Chemistry is the identity and the molecular imprint of the bioactive molecules, and without it the understanding of their biochemical behavior is impossible. The physical and therefore the biophysical approach of biomolecules and biomaterial behavior contributes to the definition of the term medicine. We can suggest the following definition on what a medicine is: *Medicine is the commodity that is composed of the bioactive molecule, known as drug (i.e., pharmacologically active), and of the biomaterials, known as excipients (without pharmacological action and toxicity) that affect the therapeutic efficacy. The excipients are rationally chosen and interact with the bioactive molecule in an extent that is defined by their physical, chemical, and biological properties, aiming to maximize the bioactive molecule effectiveness that will be used to improve or to restore the physiological functions of the human organism.* We have to take into account that the medicine in its parts (bioactive molecule and excipients) and as a final marketed product could be considered as a *biomaterial*, and apart of its chemistry, it should be studied and evaluated under the principles of physics, biophysics, thermodynamics, and mathematics, in order to maximize its efficacy and to reduce its adverse drug reactions. The research in the field of excipients is attractive and could promote scientific directions to be discussed in order to establish new regulatory guidelines in the approval process of nanomedicines (see Sect. 7.2). We suggest the introduction of a new term, the *innovative excipients*. The innovative excipients refer to the self-assembled nanostructures (liposomes etc.) with new surface properties and are able to transport and deliver bioactive molecules (i.e., drugs) to the target tissue. This term could be used as the driving force in order to construct new regulatory guidelines and to facilitate the approval process of innovative medicines (see Part III).

2.1.2 Biophysics

Biophysics, as a science, delimits in the field of nanotechnology and especially in the fields of health and medicines. Generally, biophysics is the field of physical sciences that deals with the study of physical phenomena that correlate with the structure, organization, and function of biological systems; the use of physical principles and methods into the research of phenomena of life; and the study of biological results from the influence of physical factors in the living matter.

It is a promising field of physics and offers complementary issues to chemistry, applied mathematics, engineering, and nanotechnology. The first period of biophysics development is signaled from Luigi Galvani (1737–1798) studies. In 1786 he experimentally studied the effect of static electricity in the muscles of a frog. Thomas Young delivered the theory of color vision and showed the hydrodynamic function of the heart. Julius Robert Mayer (1814–1878) pointed out the recycle of all kinds of energy in living systems: heat, solar, chemical energy, and mechanical work. Hermann von Helmholtz (1821–1894) studied the propagation speed of nerve pulses and the muscle contraction, developed the three-colored vision theory, invented the ophthalmoscope for the observation of the retina, and delivered the theory of consonance. After 1930, due to the progress of physics and new techniques (spectroscopy, X-ray diffraction, etc.), biophysics offered the world important inventions and achievements that are rewarded with Nobel Prizes in medicine, physiology, and/or chemistry. It's been 50 years of hard research, from the decoding of the DNA chain from J. Watson and F. Crick (Nobel Prize, 1962) to the sequence of human gene mapping in the beginning of the twenty-first century that changed science drastically but also the direct recipients of these achievements, the people and the society. Biophysics is subdivided to specific sectors such as molecular biophysics, cellular biophysics, and biophysics of complex systems.

Sciences that are also distinguished are the membrane biophysics, the neurobiophysics, the X-ray biophysics, the medical biophysics, the environmental biophysics, and the computational biophysics. The biophysics contributes to the study of living systems and systems made of biological components such as lipids, proteins, carbohydrates, etc. The basic objectives of biophysics are the biomolecules, the cells, and the tissues. The higher levels of life organization are the study subjects of medicinal physics and health physics. As in most interdisciplinary fields, the boundaries of biophysics are wide and vague, between other biosciences (biochemistry, molecular biology, cellular biology, biotechnology, and neurophysiology). Also, biophysics shares its boundaries with physics, chemistry, genetics, applied mathematics (control theory, information theory), micro-engineering, and nanotechnology [14].

The living matter is composed of small and big molecules. Electrons, atoms, and molecules that compose the matter are characterized by stable physical properties that are independent from their origin or their historical evolution. Every molecule that is a structural or a functional substance of organisms is called biomolecule. Complex biomolecules manifest extreme properties like metabolism, self-reproduction, self-adjustment, self-repair, development, evolution, movement, reaction, and adjustment.

Biomolecules are produced by the living organisms through their metabolism and can be classified in a scale of increasing complexity depending on their molecular weight.

Organisms receive precursor compounds through food and environment. The quality of these compounds is a very important element that defines the characteristics of more complex compounds that are the structural elements of cells and therefore tissues. The structural elements are linked to each other with covalent links producing macromolecules with a molecular weight of 10^3–10^9 Da. Their stability and their physicochemical characteristics participate in the development of stable structures of the organism (tissues, organs) and in their function. The function of the organism's structural units, like cells, tissues, organs, ensures the possibility of adjustment to the microenvironmental changes that result from external changes or pathogenic factors (viruses, bacteria, etc.).

The biomolecules are characterized from stability and diversity that is a result from the evolution process of organisms that target their survival, counteracting the second thermodynamic principle. This process led to quality characteristics through the confrontation with the microenvironment and the environment in general.

In the next chapters, we will see that the logic in biophysics, biomolecules, biomolecular structural units, and biostructures (e.g., cells) is directly involved with the development and study of nanostructures and nanosystems that are used in therapeutics and/or diagnostics. It is obvious that the above characteristics of biomolecules differentiate them from the molecules that are not elements of organisms and do not participate in structural elements, but mostly in functional properties.

In order to participate to the creation of organism's stable structures, as we mentioned above, the biomolecules need to be physicochemically stable. The elements that have been chosen for the biomolecular composition through metabolism are carbon, hydrogen, oxygen, and nitrogen. These chemical elements are biologically compatible, and their atoms have a common property, a great easiness in forming covalent links with equal electron contribution. It is known that the covalent link stability is conversely proportional to the atomic weight of the elements involved.

These elements have been chosen, so that they can create biomolecules through covalent links. The biomolecules are chemically stable and can create stable biological structures with great diversity. This diversity is directly related with the functions of biostructures and is fully connected with the phenomenon of life. At this point, we must mention that the carbon atoms can form covalent links directly with oxygen, hydrogen, nitrogen, and sulfur. This way there is a possibility of forming a great number of individual structures that can participate in complex biostructures through the metabolism of living organisms. Apart from the air, the liquid, and the solid state of matter that we meet in nature and are the most important states of medicines, there is the liquid crystalline state (see Chap. 2). The liquid crystalline state of biostructures – which consist of biomolecules like phospholipids, cholesterol, lipids, etc. and/or other nontoxic but biocompatible and biodegradable biomaterials – can be the one characterizing the modern systems of bioactive molecule transport according to the science of nanotechnology (see Chap. 5). The liquid crystalline state of nanosystems has the ability of structural biomaterial and autonomous biostructure diversity. Therefore, they are functional, stable, adjustable to microenvironmental changes, and effective, for example, they transfer the bioactive molecules to the damaged tissues.

Biomaterials are an important scientific area of nanoscale technological platforms that are applied in a wide range of nanotechnological products following quality standards and the GMP (good manufacturing practice) requirements. They are related to physical sciences, biology, engineering, physiology, and clinical sciences. Last decades' development of biomaterials was the result of evolution of physical sciences and technology, but mostly a result of understanding the interaction of materials with living organisms in cell and tissue level.

There are two predominant perspectives for the use of biomaterials. The first one suggests the passive application and inactivity of biomaterials when reacting with living matter. Every synthetic or natural material could be used on each own or as part of a system, targeting the good function of organized living systems or the replacement of functional sections of an organism. On the contrary, the second opinion that was submitted later on accepts the biomaterial action when reacting with living matter. This way, every material that is designed and used reacts and takes part in the biological and physiological function of the living systems.

The applications of biomaterials are many, and when concerning health sciences, they implicate specialized and targeted activity in tissue targets, controlled release of the bioactive molecules, tissue engineering, reproductive medicine, nanotechnology, and the field of biomimetic materials. The biomaterial classification in generations is shown in Table 2.1.

Some of today's applications of biomaterials are in specialized activity in tissue targets, in controlled release of bioactive molecules (e.g., nanosystems for bioactive molecule transport), and in tissue engineering, regeneration medicine, nanotechnology, and nanobiotechnology and, of course, as biomimetic materials. Table 2.2 presents the applications of biomaterials in medicine and pharmaceutics.

Table 2.1 Classification of biomaterials into generations

First generation	Inactive materials
Second generation	Bioactive materials
Third generation	Biodegradable materials
Fourth generation	Biofunctional materials
Fifth generation	Biocompatible materials

Table 2.2 Applications of biomaterials in medicines and pharmaceutics

Development of novel biocompatible biomaterials at the cellular and molecular level
Development of biomaterials that mimic the excellence of nature (biomimetic materials) (e.g., cobweb materials/proteins)
Development of nanobiomaterials for gene transfer in order to achieve gene therapy
Development of hybrid nanobiomaterials. Combination of composite material with cells for tissue regeneration
Development of smart nanodevices for controlled drug delivery
Development of communication systems in tissues with biomaterials that will participate in this function

At this point, we must mention that the reaction of biomaterials with biological tissues premises the absence of the following: carcinogenicity, immune response/immunogenicity, teratogenicity, and toxicity.

2.1.2.1 Biophysics and the Structure of the Eukaryotic Cell

The cells of the human organism (but also the cells of the rest of the living organisms, regardless of the complexity grade they have organization) are structured from biomolecules like proteins, lipids, genetic material, polysaccharides, etc. that have the same stability and diversity properties that have been mentioned before. These biomolecules are at the same time biomaterials because apart from the functional role they have for the cell, they are also structural components of the cells. The most important part of the cell structure is the cell membrane as it is directly connected with its structural units. The biophysical approach of biomaterial organization, functionality, and cooperativity – a term that is used to describe the thermodynamic relationship and behavior of the systems biomaterials (systems approach, see Chap. 3) – is important and suggests that the physical parameters in the biological substrate can affect its survival [7].

The role of the cell membrane is important and is related with the smooth function of the cell. It controls what substances enter and exit the cell, and at the same time, it sets boundaries to the cell keeping its shape. The membrane is composed of a double layer of amphiphilic lipids, mostly phospholipids and glycolipids. The outer side of each surface is mostly composed of ionic and polar groups that interact with water molecules (polar solvents). The inner side is composed of the lipid hydrocarbon chains. The lipid chains are placed parallel and under normal conditions there is enough flexibility allowing rotation around the alkyl chains. The nonpolar sides of the chain come together in the middle of the lipid layer forming a hydrophobic barrier that is not permeable from polar molecules but permeable to small hydrophobic moles. The intact proteins and lipid molecules that are limited in the lipid layer can be diffused on the side. When simulating the membrane as a 2D polymorphic liquid, it can be described as a dynamic system where proteins and lipids can be transported and interact with each other. The term describing the cell membrane state of matter is the liquid crystalline state (see Chap. 2), through which the phase transitions and the thermodynamics of them are evaluated. The effect of the phase transition through the system free energy change (ΔG) is responsible for its functionality and stability [10].

The phospholipid layers that compromise the cell membrane are in liquid crystalline phase and the fluidity of the system defines its functionality. The phosphate groups can be related to the metabolic energy demands since these lipids are a self-assembled system, having their own movement (the double-layer dance) across the membrane surface. This would be the ideal way for the bonds to differentiate, while the lipids would have a definitive position and mainly concerning the distribution of the building blocks in the liquid crystalline. Generally, concerning the matter, biocolloids (that are the materials that the living organisms are structured from, like nucleic acids, proteins, lipids, and, therefore, all the biological macromolecules) are structured from self-assembled liquid crystals. It is really remarkable that the deepest pursuit of life is the achievement of the appropriate liquid crystal conformation for the smooth operation of cell and organism [8].

The liquid crystalline state that we will extensively refer at the following chapters of this book is addressed to a length of 10 nm, and this suggests that the lipid layer is analyzed into structural units of that size and determines the viscosity of the membrane. These areas may correspond to smaller ones that hold in approximately 100 lipid chains. Also, the contradictory perception of the lipid masses was inspired from the weakness in mixing the ordered and non-ordered phase in the liquid membrane model. The organization of linked areas in a special phase would be a phenomenon proportionate to phase separation during filtration. There are many mechanisms involved in the floating of compact lipid masses. Their size could be between 1 and 1,000 nm (in other words, it is an intermembrane channel that eases the entry and exit of substances of different molecular weights). Also, the time proportion of the existence of the lipid compact masses is unknown. The segregated phase has been predicted from the hydrodynamic model but has not been observed.

The ordered and non-ordered phases could be characterized as "*islands*" or "*rafts*," according to the physicochemical and thermodynamic status of the environment they are in. The islands or rafts are usually due to the changes of cholesterol concentration that affect the fluidity of the cell membrane. These "*islands*" or "*rafts*" take part in the physical variability of systems' parameters that causes biophysical abnormalities that could be defined as "biophysical disease factors" (see Chap. 4) – in the level of living organisms – and are the cause of the disease development from a physical and thermodynamic point of view [3].

It is important to mention that the *islands* and *rafts* having the biophysical behavior that we previously described can be used as prototypes for the design and evaluation of the stability of innovative delivery systems of bioactive molecules. These are all described in detail in Chap. 5.

The cell membrane is primarily composed of proteins, hydrocarbons, and lipids, with the latter being classed as phospholipids, which can create a stable structure thermodynamically important for the formation and the function of the cell membrane, the lipid double layer, so that the hydrophobic part of the phospholipid does not interact with the water environment. The reason why we refer especially to biological membranes is their similarity to the nanolipidic systems (e.g., lipidic nanocarriers, liposomes) [2]. For this reason, the nanolipidic systems are considered as bio-inspired systems (see Chap. 5) because their physicochemical functions can be related to those of the colloidal-biological membrane systems. In the early of 1960s, when Bangham identified lipidic structures as artificial cell membranes, biophysics and biologists gained an innovative platform for studying the properties of biological membranes [2, 21]. The biological membranes present abilities similar to those of the colloidal systems with lipids as basic building units, like phospholipids, cholesterol, etc. (Fig. 2.1). The phospholipid hydrocarbon chains have a length of 12–24 carbons, and because of their synthesis mechanism, they always have an even number of carbons. The chains are usually parallel to each other. The conformation of the

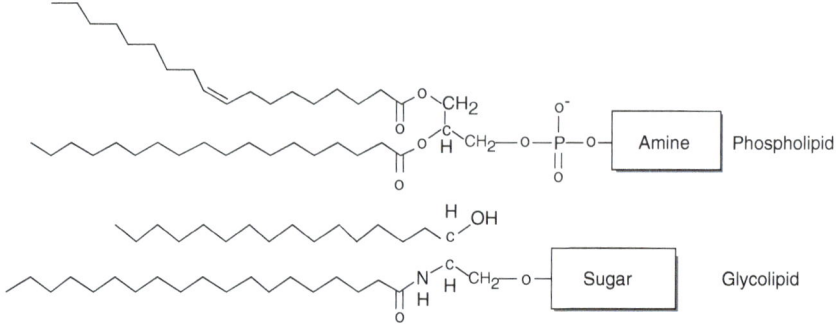

Fig. 2.1 Structure of cellular membrane's lipids

Table 2.3 Frequently occurring fatty acids in the phospholipids

Molecular structure/formula	Saturation of fatty chain	Nomenclature
$CH_3(CH_2)_{14}COOH$	16:0	Palmitic acid
$CH_3(CH_2)_{16}COOH$	18:0	Stearic acid
$CH_3(CH_2)_5 CH = CH(CH_2)_7 COOH$	16:1	Palmitoleic acid
$CH_3(CH_2)_7 CH = CH(CH_2)_7 COOH$	18:1	Oleic acid
$CH_3(CH_2)_4 CH = CHCH_2)_2 (CH_2)_6 COOH$	18:2	Linoleic acid
$CH_3CH_2(CH = CHCH_2)_3 (CH_2)_6 COOH$	18:3	Linolenic acid
$CH_3(CH_2)_4 (CH = CHCH_2)_4 (CH_2)_2 COOH$	20:4	Arachidonic acid

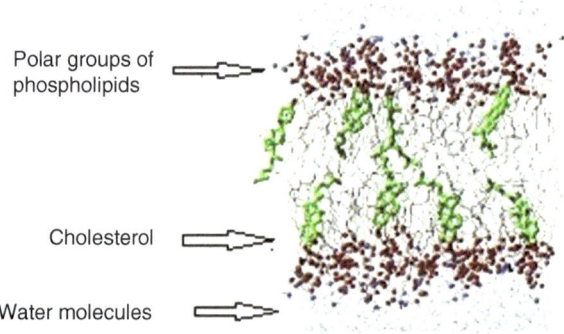

Fig. 2.2 Phospholipid cell membrane with incorporated cholesterol

carbon-carbon (C-C) bond along the chain makes the *trans* or *gauche* stereochemistry. The *trans* stereochemistry allows the chains to be close to each other, while the gauche stereochemistry widens the distance between the twisted chain and the neighbor ones. The distance of the hydrocarbon chains defines the membranes' physical abilities like lipid transfer, permeability from polar groups, and phase transition from fluid to colloidal. In colloidal phase, the lipid chains are mainly in *trans* conformation. The membrane liquid phase is characterized from high permeability of membrane components and smaller width of the lipid layer due to the *gauche* isomers of the hydrocarbon chains. The double C=C bonds in the chains can also be in *trans* or *gauche* conformation. Phospholipids in *trans* conformation have high density. On the contrary, phospholipids in *gauche* conformation, that is, the most usual state in biological membranes, have lower density due to the change of the chain arrangement. The presence of double bonds makes the length of the hydrocarbon chain smaller, and the phospholipid membranes of the unsaturated lipids are smaller compared to the ones of saturated having the same carbon atoms. The most usual lipids in phospholipid molecules are gathered in Table 2.3. The phospholipid double layers contain also cholesterol (Fig. 2.2). The membrane permeability is affected from the presence of cholesterol. The cholesterol concentration has been genetically defined in human species and is ideal for the expression of the biological function. The presence of cholesterol in the lipid membrane allows the thermodynamic organization of the double layer, making it fluid, offering its morphological and functional characteristics, and therefore keeping it alive (Fig. 2.2). The phospholipid mobility in membranes is their characteristic and is affected from their properties. The type of the phospholipid, the polarized head, the type and the lipid saturation of their chains play an important role in their mobility inside the membrane. Simulating synthetic membrane prototypes with phospholipid composition with or without cholesterol, we can have a completed picture of the behavior of a lipid system for drug delivery. The knowledge of the lipid mobility in the membranes allows thermodynamic measurements and design that corresponds to the physicochemical characteristics of the bioactive molecules that we want to enclose in them. Completing our report in the meaning of biomolecules, biomaterials, biostructures, through the science of biophysics and describing the composition, the organization and the structure of the biological systems, their stability and their complexity, but mostly their functionality, we must understand that this knowledge and these approaches have value when they can be used as tools for the development of useful systems for the protection of the human health [14].

Therefore, the laws and the principles of biophysics through thermodynamics and biomaterials are the tools to the rational design of nanosystems with applications in therapeutics, diagnostics, and imaging and for the understanding of their function and behavior in the biological environment.

It is extremely important for these senses to be transferred in the level of designing and developing nanosystems that will follow the principles of the cell in behavior, stability, diversity, and functionality. The question that follows refers to the ability of nanotechnology to develop and study the nanosystems that will be similar in composition and behavior of those of a cell of an organism. The ability of designing nanosystems similar to the cells is extremely difficult and perhaps impossible, as the cells are a result of the evolution process in billion year's time.

Having in mind the basic principles that characterize the biophysical behavior of a cell, we can search for the structural components, like lipids, cholesterol, phospholipids, etc., which have high cooperativity among them. Therefore, they will behave as a whole, according to physics and thermodynamics, and they will be nontoxic, biocompatible, and biodegradable, so that we will be able to develop nanosystems whose biophysical behavior will resemble the one of the living cells. The relationship regarding the biophysical behavior between biomembranes and cell biology has promoted nanoparticulate systems as leading nanosystems to deliver drugs to the target tissues (see Chap. 5).

As an example of the biophysical approach in the development of nanosystems, we can refer to the phase transitions of the liquid crystalline state that is present in the cells and correlate with their functionality, stability, and effectiveness. Therefore, nanosystems like liposomes, dendrimers, nanoemulsions, polymeric nanoparticles, etc. could transfer and release the bioactive molecules into the cells of damaged tissues in respect to the "decoding" of the hidden to them and the science of nanotechnology by simulating and mimicking the function and the structure of cells: this is referred to the "silence functionality."

The liquid crystalline state, the study of the nanosystem thermodynamics as it is quantified from the physics laws, and the study of their thermotropic behavior as it is known from the functionality of nanosystems in correlation to their physicochemical characteristics – for example, the study and the evaluation of size and distribution of sizes in colloidal dispersions in reference to the biocolloidal dispersions of living organisms – are very important physical approaches in pharmaceutics and comply with the science of biophysics.

From all of the above, we conclude that we can build nanostructures and nanosystems, study them, and evaluate them making nanotechnological products that can be used in disease diagnosing, damage tissue imaging, and their cure.

Before mentioning the nanoparticles that belong to soft matter (see Chap. 5) and the biomaterials that are used to formulate them, we must mention the basic principles and properties of the state of the matter that is related with our biological cell, the liquid crystalline state. The liquid crystalline state of matter and its transition phases ("metastable phases"), which we will mention later on, consist of the basic knowledge into understanding the phenomena of life and the nanosystem design and development, exceeding the biochemistry knowledge for systems and presenting them by their biophysical approach.

2.1.2.2 Liquid Crystalline State of Matter

Liquid crystals are thermotropic (Fig. 2.3) in that their transition state depends on the temperature, while liquid crystals are lyotropic in that their transition state depends on the temperature and their concentration in the medium that they are found. The basic difference between the thermotropic and lyotropic liquid crystals is that the anisotropic dispersion forces between structural units are responsible for the thermotropic transitions (intermediate states with different structural unit orientation of the thermotropic liquid crystals), while the dispersion media are primarily responsible for the lyotropic transition of the liquid crystals (and secondly is the temperature change).

Enantiotropic are the thermotropic liquid crystals that transit to liquid crystalline state either by lowering the temperature of a liquid or by raising the temperature of a

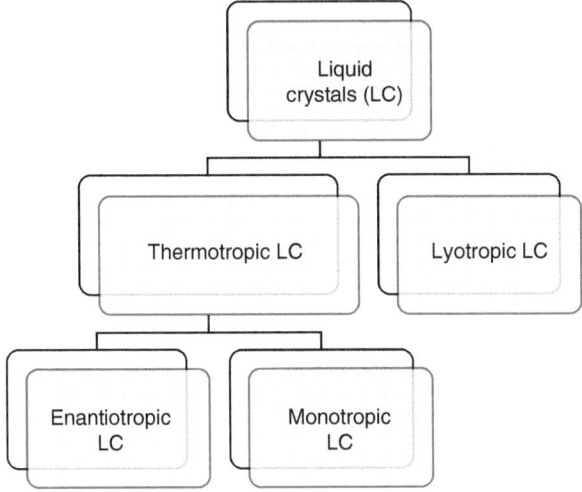

Fig. 2.3 Classification of liquid crystals

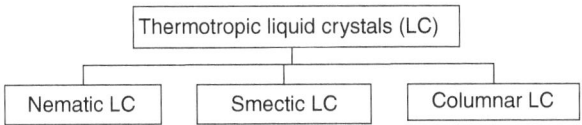

Fig. 2.4 Classification of thermotropic liquid crystals

solid. Monotropic are the thermotropic liquid crystals that transit into the liquid crystalline state only by lowering the temperature of a liquid or only by raising the temperature of a solid, not with both ways at the same time (main difference with the enantiotropic). The thermotropic liquid crystals are classified as shown in Fig. 2.4.

The thermotropic liquid crystals are classified as nematic, smectic, and columnar. In nematic liquid crystals (Fig. 2.5), the molecules (structural units) are free to move to all directions, are parallel without setup, and move along the direction axis. In smectic liquid crystals (Fig. 2.6), the molecules present a transition grade, set up along one direction, and have the tendency to line up in layers or in plateaus. Finally, in columnar liquid crystals (Fig. 2.7), the molecules are in disk shape (and not as linear rods/bars) forming piles. Each one of the categories of thermotropic liquid crystals that were mentioned before presents new transition phases (that are subdivisions of the above classification) according to the molecule arrangement. For example, according to the degrees of molecule assembly and the place they take in each level, we can discriminate 12 smectic phases.

The lyotropic liquid crystals consist of one polar and one nonpolar part. The polar part is an ionized chemical group, while the nonpolar is a hydrocarbon (saturated or unsaturated) chain. The lyotropic liquid crystals are surfactants due to their chemistry (while the opposite doesn't always apply, not all surfactants form liquid

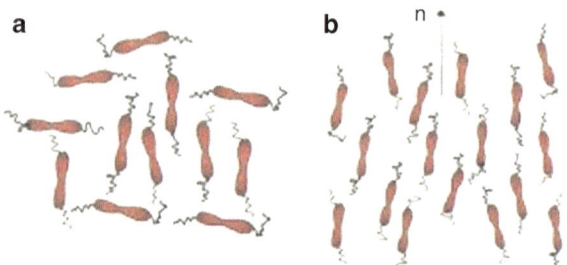

Fig. 2.5 Molecular order of nematic liquid crystals. Schematic representation of the instant assembly of rodlike molecules into the isotropic phase (**a**) and in the polar uniaxial nematic phase (**b**). In both figures, the positions of the molecules are random, while in (**b**) the addresses of molecules show statistical alignment to direction/directional *n*

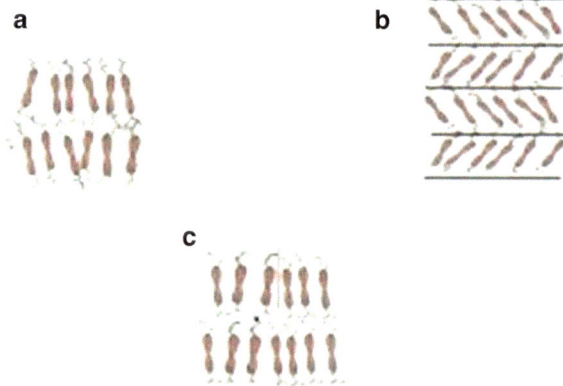

Fig. 2.6 Molecular order of smectic liquid crystals. (**a**) Smectic A (SmA): complete disorder of molecular positions within the smectic layers. (**b**) Smectic C (SmC): complete disorder of molecular positions within the smectic layers. (**c**) Smectic B (SmB): partial order of molecular positions within the smectic layers

crystals). The lyotropic liquid crystals are called this way because they are formed due to the mixing of two or more, different to each other, components, not being in liquid crystalline phase on their own. There are two basic categories of lyotropic liquid crystals:

- Solutions of anisotropic colloidal macromolecules or particles whose structure presents powerful orientation (e.g., rigid or semirigid polymer bars). The role of the dispersion media in this case is to form the appropriate dilution, so that the liquid crystal fluidity will appear, while the order (e.g., nematic, smectic, etc.) in the direction will be ensured from the orientation of the macromolecular structure, therefore, from the material itself.
- Nonliquid crystalline molecules in solutions that are self-assembled in supramolecular structures that are self-organized in liquid crystals. They are usually formed

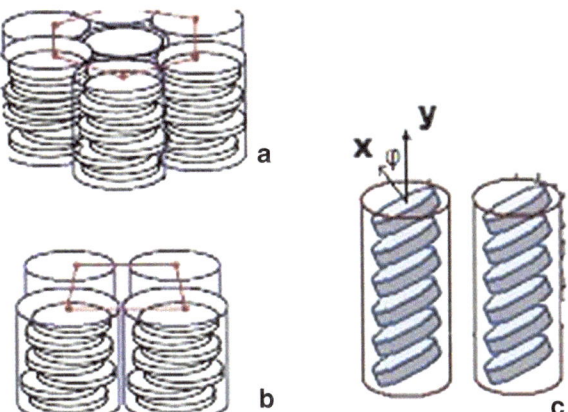

Fig. 2.7 Molecular order of columnar liquid crystals. (**a**) Columnar hexagonal phase: hexagonal network/grid of columns in the level x, y. (**b**) Columnar orthogonal phase: orthogonal network/grid of columns in the level x, y. (**c**) Columnar inclined phase: the directional deviation from the angle by column axis by angle φ

from the dilution of amphiphilic molecules in the appropriate solvent. Changing the component concentration, the solution goes through different liquid crystalline phases that according to symmetry and order/disorder are proportional to the ones in thermotropic phases, nematic, smectic, and columnar.

The entities that follow this order can be either molecules or some supramolecular units (micelles) that result from the molecular self-assembly. For example, we have the smectic phase in which the molecules form layers, while the formation mechanism of the columnar phase has two stages: the formation (self-organization) of supramolecular tubes and the self-organization of these tubes in columnar phase. The dependence mechanism of the molecular self-assembly is led from the tendency of the hydrophobic units of the amphiphilic molecules to decrease the interface that is formed with the solvent and proportionally, from the tendency of the hydrophilic ones to increase it. This phenomenon is concentration dependent. As it has already been mentioned, for the liquid crystalline transition phase apart from the temperature, the concentration is also a main factor. Depending on the temperature, the amphiphilic molecules can form micelles or liquid crystals. Right before the temperature of the shape formation, the solution remains clear. Also, the amphiphilic molecule organization in solvent, like water, depends on the particle concentration in the dispersion.

More in particular, in low concentrations, the molecules are randomly dispersed in the medium. In high concentrations, on the contrary, the hydrophilic heads are intact with the water, while the hydrophobic hydrocarbon chains are oriented in a way where they do not come in touch with the solvent molecules. Raising the concentration in a solution of anisotropic polymers or colloids, we can observe transition from isotropic phase (random orientation) to nematic phase (the molecules are oriented without having crystalline 3D structure). When the concentration rises enough, the molecules begin to self-organize themselves in micelles that have different shapes. Usually they take the shape of a sphere that is empty on the inside.

The polar groups are at the outside of the micelle while the hydrophobic tails are on the inside. The size of the micelles depends on the molecular concentration (they enlarge as the molecular concentration increases). According to the micelle size, the dispersion is either clear or it fogs (forming microemulsions or emulsions). It is important to mention that the membrane lipid double layers of the cells act as lyotropic liquid crystals. This observation is important as we can design nanosystems that can be used in therapy or as cosmetics with increased efficacy [1].

2.1.2.3   Liquid Crystals in Biological Systems

The cell membrane of living entities is a natural liquid crystal. Its structure is a lipid double layer with phospholipids as dominant components. In both sides of the double layer and on the inside, there are proteins and other biomolecules. The harmony (Greek word is $\alpha\rho\mu o\nu i\alpha$) in the coexistence of the molecular order and the fluidity of the liquid crystal phase gives the cell membrane the required mechanical stability, permeability, and general functionality. This way the cell membrane functions as a proper cover of the cell contents and offers the ability of molecular movement so that the physicochemical processes would have the required speed that is needed for the interaction with the environment.

The phospholipids are amphiphilic molecules that are composed from a polar head groups and two hydrocarbon chains (tails). In the cell membrane of the animal kingdom species, there are only a few phospholipid types that are different in the chemical structure of the polar part, while the length of the hydrocarbon chains varies (16–20 carbon atoms).

The particular structure of the membrane is defined from the general characteristics of the phospholipid molecular structure. The structure of the polar head defines the chemical processes that take part on the surface of the membrane. The length of the lipid chain and their relative flexibility control the temperature limits of the membrane crystalline state. In lower temperatures, the membrane transits from the liquid crystalline phase to the gel phase. In the gel phase, the flexibility of the lipid chains decreases and the polar heads orient themselves in a hexagon mesh. As a result, the membrane actively restricts the interaction of the inside of the cell with its environment. It has been observed that the temperature that cells in the living organisms function is slightly higher than the transition temperature (see Sect. 2.3) of cell membrane phospholipids from the liquid crystalline phase to the gel phase.

The lyotropic liquid crystal nanostructures are the dominant structures in living organisms and their part on the cell function and preservation of life is definitive. The contribution of the lyotropic structures in organization and fluidity of the cell membrane but also in the preservation of homeostasis and in matter exchange is characterized as definitive for the evolution process of the living organisms. Also the lyotropic liquid crystals have a special interest in the field of biomimetic chemistry. The components of the biological membranes – i.e., phospholipids, lipids, cholesterol, etc. – are structurally and thermodynamically organized so that the membranes are fluid and flexible. The biomolecules that compose the membranes can move into the fluid mosaic without leaving the cell. They can move freely into the membrane, changing orientation and direction, and diffuse into the membranes (*flip-flop* and *lateral diffusion* phenomena) [16]. These membranes that are in liquid crystalline state have on their inside macromolecules, like protein receptors and other surface macromolecules that participate in the organization according to the absolute cooperativity with the main structural elements of the membrane. There are also many biological structures that have the liquid crystalline behavior like the concentrated colloidal protein dispersion that comes out from the spider to produce its web. The sequence of the biomolecules on the web is critical to get the necessary durability. DNA and polypeptides can be in liquid crystalline phases, and it must be mentioned that the biological transition is most of the times chiral and plays an important role in their structure and function. There are many applications of the lyotropic liquid crystals in pharmaceutics. A lyotropic liquid crystal can be used to cover up a bioactive molecule and protect it along the gastrointestinal tract. As a result, the bioactive molecule can be taken orally, and when it reaches the target, it is released as the liquid crystal dissolves. Lipids can be organized in biostructures and biosystems forming colloidal dispersions into the organism for the enclosure and transport of hydrophilic and hydrophobic substances of the living organisms. The predominant membrane lipids can give scientists new ideas for the development of lipid nanosystems for the bioactive molecule transport in cells and target tissues. Anticancer and antifungal medicines, vaccines, and other pharmaceutical products are being administered with lipid carriers that transfer them to the target tissues.

The predominant lipids that we meet in cell membranes are the phospholipids, whose hydrophobic section consists of long chains of saturated or unsaturated fatty acids. Their amphiphilic properties are of crucial importance, as they take the appropriate orientation when in water environment, forming lipid double layers in the logic of favorable thermodynamic arrangement of the fatty acids chains, the development of hydrophobic interactions, and the minimization of the system free energy. In water environment, phospholipids are in very distinct transition phase known as L_α, L_β, and $L_{\beta'}$. The capital letter (L) corresponds to the characterization of the system structure-organization (lamellar or leaf structure) (long order), while as a subscript, the letter of the Greek alphabet corresponds to the characterization of the system structural unit configuration ("$_{\beta'}$" corresponds to the system structural unit configuration in a specific angle) (short order).

The L_β phospholipid transition structure concerns the lamellar gel phase, at zero angle structural unit (phospholipid) configuration. The L_α transition state concerns the liquid crystalline state. The phospholipids have the ability of transit from the gel phase to liquid crystalline state by temperature rise. In this case, the phospholipid molecules move in greater speed increasing the lateral motion and diffusion and the *flip-flop* motion and diffusion between them, while the intermolecular movement around the C-C bonds increases, forming knocks.

The phenomenon of conformational polymorphism plays an important role as it affects the physical properties of phospholipid transition phases, and as a result, the release and the pharmacokinetic profile of the bioactive molecule that is encapsulated into the nanosystem are changed, if the phospholipids are the structural units of such nanosystem (e.g., liposomes, Chap. 4).

Referring to cell membrane transition phases, we can notice that they have an important side of their properties and they are correlated directly to their structure. The temperature variations lead to cell liquid crystalline state transitions. These could be due to physiological variations of the biological environment physical chemistry and are important for the organism homeostasis. According to the cell membrane behavior, we can assume that the control of the biomaterials' liquid crystal state and transition state has important advantages into the enclosure and transport of the bioactive molecules in the organism.

It must be mentioned that the biomolecule structural characteristics (i.e., the double bond number, the fluidity rate, the length of the fatty acid chain, etc.) that contribute to the cell structure and the structure of a nanosystem that transfers bioactive molecules affect the transition temperature that is related to the biomolecules' orientation (i.e., phospholipids, lipids, etc.) and the speed of diffusion phenomena, into the membrane. This way the increase in the saturation of the fatty acids lowers the transition phase temperature. The length and the hydrocarbon bonds change in the chains, affecting the degree that the lipid molecules can be compressed and, therefore, the membrane fluidity [10].

Indeed, the unicellular organism temperature changes according to their environment. They change their metabolic path for production of the necessary biomaterials by adapting to the temperature changes and therefore preserving the fluidity and the functionality of the cell membrane. It has been observed that the cold-blooded animals have a greater percentage of unsaturated chains in their cell membranes than the warm blooded. It is possible that the unsaturated chains give a greater stability in liquid crystalline phase as the temperature changes. The presence and the cholesterol concentration percentage also affect the coherence and the fluidity of the cell membrane by participating in the organism homeostasis and the function of the cell's basic unit.

2.1.2.4   Properties of Liquid Crystals

Liquid crystals are extremely complicated and their elongated structure allows the light to be transmitted in different speed, through their structural units. This leads to a dipole reflection that describes this unusual behavior. This insight allows us to study in detail the liquid crystals through crossed polarizers under the microscope and is the foundation for most applications. In crossed polarizers, a usual, common fluid does not change when placed between them, as the liquid doesn't affect the light polarization as it will be absorbed from the second polarizer. On the contrary, placing a liquid crystal material between crossed polarizers, we can observe that the polarized light turns in a way. As the light turns through the liquid crystal, the two polarizers travel toward the same direction, but in different speed. As the two beams of polarized light travel, one is ahead of the other due to their property to double reflect. The two polarized beams emerge again from the specimen in different phases.

The light emerging from the liquid crystal specimen is elliptically polarized, allowing a quantity of light to emerge from the second polarize. While the liquid crystal materials change their visual image, when trying a transition from one phase to another, during that exact phase, can become ideal materials for thermotropic applications. Choosing the appropriate structural units for the liquid crystal production and thermometer production, we can broaden the temperature changes. On the contrary, in liquid crystal monitors that are based on the nematic phase properties, temperatures beyond classic limits are a necessity [1].

Materials for the liquid crystal production in nematic phase are now available in temperatures between −30 and 90 °C in special applications that can be placed in variable temperatures.

A characteristic property of the liquid crystals is their behavior in the electric field. This behavior is related to the molecular charge of the liquid crystal. The liquid crystal molecule is inherently consisted of a positively charged and a negatively charged molecule. In the presence of an electric field, the charged sides of the molecule try to resist the forces and are set along the electric field direction. If the liquid crystal molecules are not sensitive to this kind of differentiation during charging, the electric field can move the positive charge at one side and the negative at the other side, forming a dipole. This has as a result the arrangement of the liquid crystal molecules in the electric field. Generally, the liquid crystal molecules can have either permanent or temporary dipoles that are responsible for their arrangement in the electric field. Most of the liquid crystal molecules correspond with the field by arranging their longitudinal axis across the electric field.

Liquid crystal molecules behave in a similar way in magnetic fields. When there is a magnetic field, some of the charges within the molecule act like very small mobile chargers. This is called induced magnetic dipole with an orientation of the North and South Pole along the direction of the magnetic field. The induced magnetic dipoles, along or transverse the elongated direction, possibly give to the liquid crystal molecules a direction either parallel or vertical to the magnetic field, accordingly. The magnetic fields are most of the times inappropriate and unpractical for this kind of applications and generally they are used for research purposes.

2.2 Nanocolloidal Dispersion Systems

2.2.1 Introduction

Disperse systems can be classified according to particle size in molecular, colloidal, and coarse dispersions. In colloidal dispersions, dispersion phase particles are in sizes between 1 and 500 nm, in coarse dispersions exceed 500 nm, and in molecular are less than 1 nm size [9].

How important are colloidal systems in everyday life? If they are, it is interesting to study how they affect life sciences and how we can use our knowledge to improve various parameters in our life, like in health. Life sciences are related directly or indirectly with biological systems. Sciences like physics, mathematics, chemistry, and science of materials, engineering, as well as any of their combinations can improve the quality of the agricultural products and the environment and affect the biotechnology and biological processes. The heterogeneity of living systems is well established, referring to macromolecules like proteins, polysaccharides, amphiphilic molecules, etc. that exist in water, while other life processes take place in the interface of these systems. Membrane's structural molecules have a specific function, e.g., cell wall glycoproteins are related to aggregation and cell growth, while the mononuclear phagocyte system (MPS) takes part in phagocytosis. The protein aggregation or agglomeration process relates to various diseases, for example, Alzheimer's disease is related to the aggregation of amyloids. Biological systems contact biomaterials, e.g., for pharmaceutical applications like controlled drug release systems produced by biomaterials (monoclonal antibodies, macromolecules) in order to target damaged tissues, etc. We can say that biomaterials and other biological components interact with biologic systems surfaces developing interfacial interactions. Biomaterial adsorption into cell affects their biological functionality. Pharmaceutical products in colloidal shape can affect various system functions possibly improving them in human favor. Pharmaceutical industry tries to develop high-quality colloidal biomaterials, emphasizing in controlling their properties so when in the appropriate environment, they will influence it to the desired direction. We highlight that colloidal are the systems that the dispersed particles are in sizes between a few nanometers to a few thousands nanometers. When referring in dispersed particles of a few nanometers, at first sight, it looks like they act like molecular solutions. These solutions have completely different properties when compared with the properties of solutions that have particles in molecular size at the same concentrations (e.g., decrease of osmotic pressure, degradation of melting point, etc.). The upper limit of the particle size is the point where (Brownian motion) thermal movement allows and reserves the particle in dispersion state. The attractive interactions between the particles are minimum and proportional to their masses. A colloidal system acts according to the equilibrium state between thermal motion and interaction forces between particles. The colloidal systems dispersed particles have great surface per volume unit. This way, the interface effects can be interpreted mainly according to the surface tension (γ) ("γ" is the force opposing to the surface growth due to attraction of water molecule or dispersed particles to the inner fluid due to interactions) and the ζ-potential [19]. Due to the great surface per volume unit, the system has great surface energy that is expressed from the following equation:

$$E = \gamma S \qquad\qquad (2.1)$$

where γ is the surface tension and S is the particles' surface.

At this point, the particles' total surface area in colloidal solutions must be mentioned. In one liter of colloidal solution or better colloidal dispersion where the volume of colloidal particles with a diameter of 10 nm is 1 %, the correspondent interface area is 6,000 m^2. This corresponds to a low surface area (S) of the colloidal system which could be achieved by aggregation or agglomeration or a mechanism that lead to entropy increase. If the aggregation phenomenon is prevented with the addition of an electrolyte – or the agglomeration phenomenon is prevented with the particles ζ-potential increase – then the surface tension will decrease and therefore achieving system free energy (G) increase. Figure 2.8 presents classification of colloidal systems. Table 2.4 presents the classification of colloidal dispersion systems.

The approach to be followed for developing an appropriate nanocolloidal system is crucial in order to formulate bioactive molecules (small compounds) or therapeutic biological substances like peptides, proteins, genes, and oligonucleotides. Also, in categories of bioactive molecules that are used to treat diseases like cancer (anthracyclines) and fungal infections (polyene antibiotics, amphotericin B), nanocolloidal systems for substance transfer are proven more effective especially in reducing their toxicity.

ζ-potential is an important physical parameter which corresponds to the total charge of nanoparticles' slipping plane (ζ-potential is not the surface charge of nanoparticles) and how it is affected from the physicochemical characteristic changes

Fig. 2.8 Classification of colloidal dispersion systems

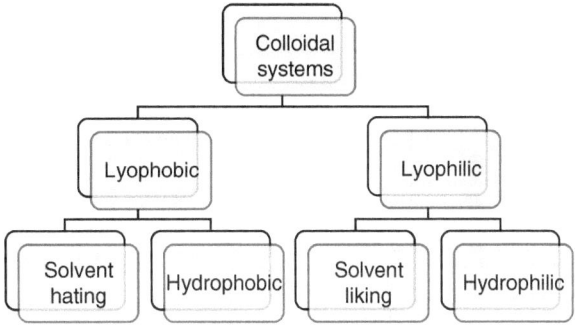

Table 2.4 Colloidal dispersion systems

Dispersion medium	Solid, S	Liquid, L	Gas, G
Solid	Solid suspension (bone wood, various composite materials)	Emulsion (opals, pearls)	Foam (loofah, bread, styrofoam)
Liquid	Liquid suspension (blood, polymer latex, ink)	Emulsion (milk, rubber, crude oil, shampoo)	Foam (detergent foam, beer foam)
Gas	Aerosol (smokes, dust aerosols)	Aerosol (fog, sprays)	–
Biopolymers	Biological media	–	–
Hydrophilic gels	Chromatography materials	Gelatin, celluloses	–
Conjugated colloids	Surfactants, microemulsions, biological membranes, liposomes	–	–

of the nanoparticle environment (e.g., electrolytes, pH). ζ-potential will be thoroughly discussed later on, but it is a physical parameter that can offer information related to the interaction of nanoparticles between them or with cells and information on the nanosystem stability under certain conditions.

Colloidal particles (nanospheres, nanocapsules, liposomes, etc.) can be administrated intravenously, orally, transdermally, subcutaneously, and intramuscularly. The colloidal nanoparticle intravenous administration is the ideal route of administration, despite the problems appearing, like the activation of the MPS, the possible lack of particles' physicochemical characteristic control (e.g., size and size distribution), their toxicity, and the vascular endothelial quality in target areas [18].

Per os is the most common way for drug administration. This route presents difficulties caused by the successive barriers that the bioactive molecules meet and by the bioactive molecules, themselves. It is the most beneficial route of administration offering low cost and patient compliance.

The colloidal system study for applications in biological and pharmaceutical sciences has presented significant progress during the past 30 years. Nowadays researchers emphasize in studying the steric effects of sterically stabilized nanosystems, i.e., invisible from MPS or Stealth™ (trademark of Liposome Technology Inc.) bioactive molecule transfer nanosystems. Also targeting damaged tissues is an important research field either by monoclonal antibodies or by molecules recognizing specific receptors in the cell surface like sugars, lectins, growth factors, etc. Also, biotechnology products that are sensitive due to their disintegration into biological fluids (genes, oligonucleotides, proteins, peptides) are trapped and transferred in nanosystems, improving even their cell entrance.

Emulsions and suspensions are dispersion systems, where a liquid or a solid phase, respectively, is dispersed in an outer liquid phase. The continuous phase is the one where the dispersed phase is distributed. Emulsions and suspensions are intrinsically thermodynamic systems and need stabilizers to ensure their useful lifetime. Pharmaceutical emulsions and suspensions are in colloidal state, i.e., the dispersed phase particles have a size of few nanometers to few micrometers (visible). Suspensions may have either a water or oil continuous phase. Aerosols are dispersed systems of a liquid or solid in gas state.

The forces that appear among particle and nanocolloidal system interaction are:

- *van der Walls* interactions or electromagnetic forces (attractive forces)
- Electrostatic forces (repulsive forces)
- Born forces – mostly low range (repulsive forces)
- Steric forces (repulsive forces) due to adsorbed molecules (especially macromolecules) in the particles surface
- Solvation hydration forces

Lastly, the use of these systems in diagnostic methods is under development, and these systems with targeting properties, as mentioned above, are useful tools for ex vivo cell therapy.

2.2.2 Stability of Nanocolloidal Dispersion System: DLVO Theory

Nanocolloidal dispersion system stability is shown in Figs. 2.9 and 2.10. The stability of a colloidal system is related to the difference in density between the dispersed particles (ρ_1) and the media (ρ_2) of dispersion, the particle size (α), and the viscosity (n). According to Stokes' law, the rate of sedimentation or creaming u of a globular particle in a liquid media, having a viscosity n, is according to the following equation:

$$\upsilon = \frac{2g\alpha^2 \left(\rho_1 - \rho_2\right)}{9n} \tag{2.2}$$

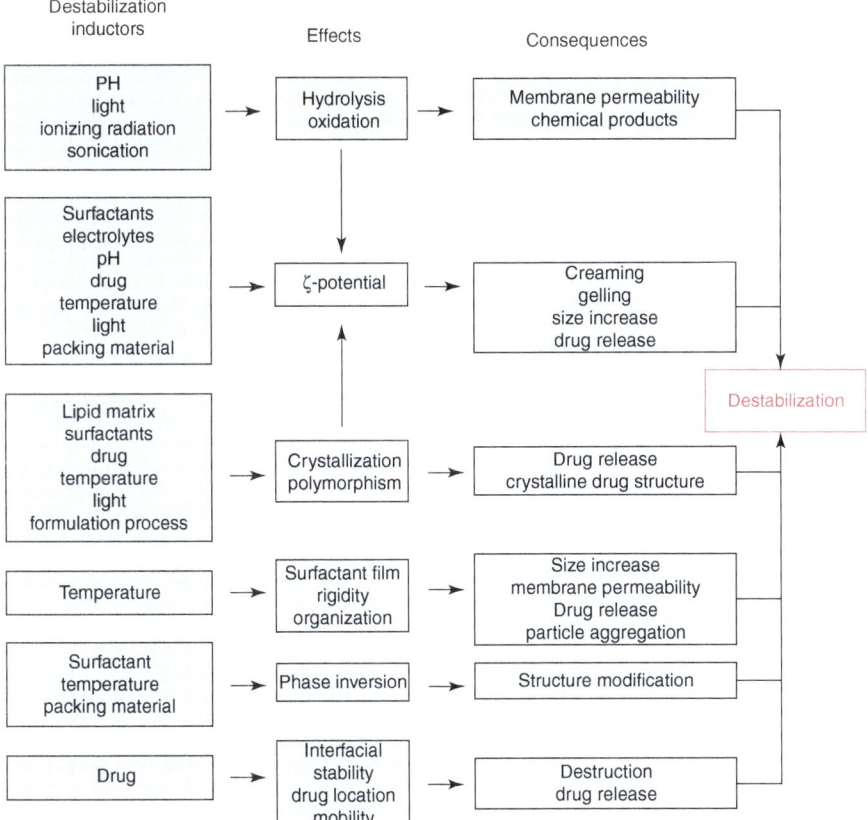

Fig. 2.9 Destabilization mechanisms and consequences on colloidal dispersion systems (Adapted from [11] with permission from Elsevier)

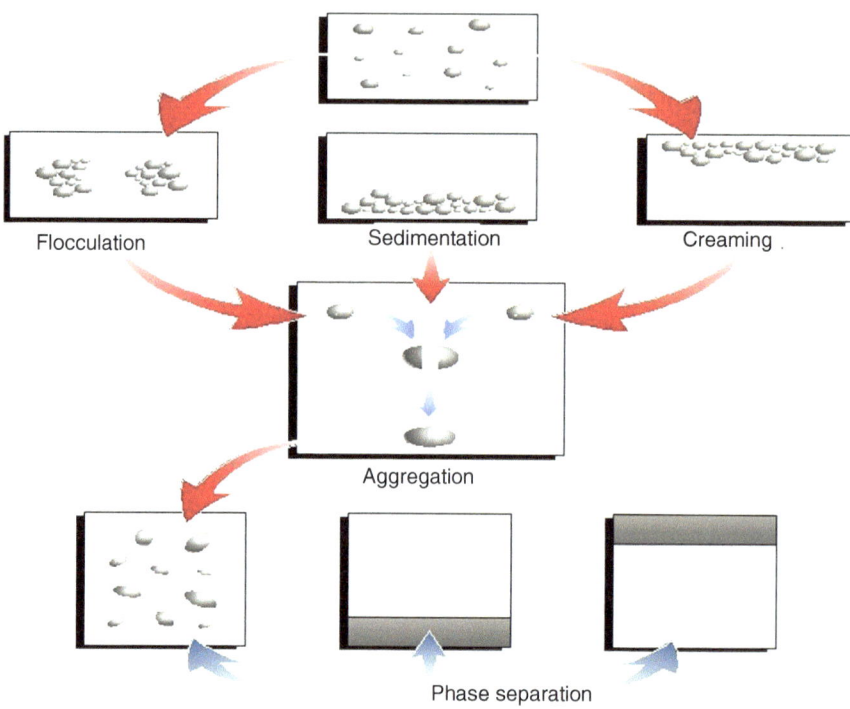

Fig. 2.10 Schematic configuration of the destabilization of a colloidal dispersion system

where α is the particle radius, ρ_1 is the particle density, ρ_2 is the media density, and g is the gravity standard.

Nanocolloidal dispersion system stability is directly related with DLVO theory. In 1940, having in mind the forces mentioned earlier and mainly the electrostatic (repulsive) and the *van der Waals* (attractive) forces, scientists Derjaguin, Landau, Verwey, and Overbeek (DLVO) suggested a satisfactory theory which approaches the aggregation process of particles dispersed in aqueous media. This theory is known as DLVO theory.

Debye and Huckel in 1923 described a theory in charge distribution in ionic solutions. Later on, Levine and Dube noted that Debye and Huckel's theory is applicable in colloidal dispersion systems. Levine and Dube's contribution was that they discovered strong medium-range repulsive forces and weaker attracting forces of high range between charged colloidal particles. We have to notice that the theory did not provide explanations on the physical instability of colloidal dispersion systems through irreversible aggregation in high ionic strength solutions. Derjaguin and Landau in 1941 [6] developed a theory related to physical stability of colloidal dispersion systems that interprets the instability effect due to strong but short-range *van der Waals* attractive forces. Verwey and Overbeek, in 1948 [23], independently made the same observations and they concluded to the same result. DLVO theory

(Derjaguin- Landau-Verwey-Overbeek) was the appropriate one to fill in the gap of the Levine and Dube theory, concerning the colloidal dispersion systems physical stability related to electrolyte ionic strength.

Classic DLVO theory refers to all forces developed between the two charged particles that are in the same dispersed media having a specific distance between them (Fig. 2.11). All forces are expressed as the total of *van der Waals* attractive forces and repulsive electrostatic interactions. *van der Waals* interactions (V_A) are attractive forces developed when two particles of the same radius r have a specific distance x between polar and nonpolar molecules. These forces do not appear when the distance between the particles is more than x and depend on the Hamaker constant (A) that is the attractive parameter related to particle chemical composition. According to particle shape, the attractive potential (V_A) ($_A$: attractive) is calculated by the following equation:

$$V_A = -\frac{Ar}{12x} \tag{2.3}$$

(V_A: *van der Waals* attracting interactions). This equation is referred to spherical-shaped particles, where A is the Hamaker constant, r is the radius of the spherical particle, and x is the distance between particles.

$$V_A = -\frac{A}{12\pi x^2} \tag{2.4}$$

(V_A: *van der Waals* attracting interactions). This equation is referred to flat-shaped particles, where A is the Hamaker constant, r is the radius of the spherical particle, x is the distance between particles, and $\pi = 3.14$.

The electrostatic repulsive forces (V_R) ($_R$: repulsive) are present only in colloidal systems where the dispersed particles have surface charge (potential). Also, they depend on ionic power (I), the pH of the dispersion media and its dielectric properties, and on the surrounded electric double layer that covers each one of these two particles. The electrostatic repulsion is the main force preventing aggregation or agglomeration

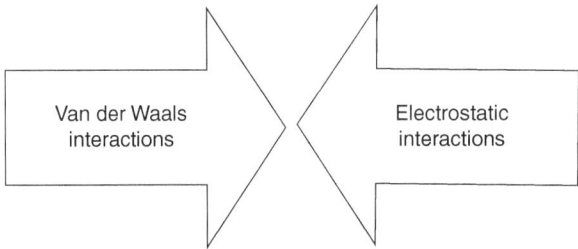

Fig. 2.11 Interaction between particles according to the classical DLVO theory

development in a colloidal dispersion system of charged molecules and is given from the following equation:

$$V_R = \frac{8K_B^2 T^2 \varepsilon r \alpha}{e^2 z^2} e^{-\kappa x} \left[\frac{e^{\frac{ze\psi}{z\kappa_B T}} - 1}{e^{\frac{ze\psi}{z\kappa_B T}} + 1} \right]^2 \tag{2.5}$$

This equation is referred to spherical particles, where κ is the Debye length, r is the radius of the spherical particle, K_B is the Boltzmann constant, T is the absolute temperature, ε is the electric constant, ψ is the surface potential, z is the chemical potential of the counterion, α is the chemical constant, n is the viscosity in electrolyte solution, and x is the distance between particles.

$$V_R = \frac{64 n K_B T}{\kappa} e^{-\kappa x} \tag{2.6}$$

This equation is referred to lamellar particles, where κ is the Debye length, T is the absolute temperature, K_B is the Boltzmann constant, n is the viscosity in electrolyte solution, and x is the distance between particles.

After the presentation of the interactions taking place in the DLVO theory, the following equation describes the total potential V_T (T: total):

$$V_T = V_A + V_R \tag{2.7}$$

Like each of the interactions separately depends on and is affected from some parameters, the total dynamic V_T is related to:

- Polyelectrolyte concentration
- pH of the dispersion media
- The valence of the metals that can be added to the colloidal system

According to DLVO theory, depending on which of the two kinds of forces will dominate in a colloidal dispersion system, there are two different cases:

- $V_A > V_R$: the attractive forces are predominant, and therefore, the particles come together, creating aggregates or agglomerates and the system that is characterized from thermodynamic instability and collapses through processes like sedimentation, agglomeration, clouding, and creaming.
- $V_R > V_A$: the repulsive forces are predominant, and therefore, the majority of the particles are away from each other, resulting to a thermodynamic and, therefore, colloidal stability.

During the period between 1940 and 1948, the DLVO theory was related to colloidal dispersion systems like emulsions and suspensions, and the description of these properties was through this theory that was adequate at that time. A naturally occurring colloidal dispersion system is the human milk that is mostly consisted of lipid dispersions. These lipid dispersions that can be found in lipid colloidal dispersion systems cannot be considered as well-defined self-assembled structures, e.g., liposomal structures. Liposomes, as mentioned in the corresponding chapter (Chap. 4), demand energy for their development in order to maintain their effectiveness besides their thermodynamic instability. In 1965, when Bangham suggested the term "liposome" (from the Greek word λίπος=fat + σώμα=body) to the scientific society, liposomes were concerned as colloidal dispersions in *nano*-dimensions that developed after experimental processes, and DLVO theory was capable for explaining the liposomal stability as a lipidic colloidal dispersion systems [20].

The weakness in interpreting effects with classic DLVO theory developed the need to extend this theory. The forces developed between particles in a colloidal dispersion system were the driving force for extended DLVO theory. Scientific efforts have contributed in discovering and measuring forces that take place during the organization of such a colloidal dispersion system. Apart from the electrostatic and *van der Waals* forces, the additional interactions that have been mentioned could be hydration forces, hydrophobic interactions, oscillations, forces related to various membrane shapes, or interactions related to water network structure.

Important variations between experimental results and theoretical predictions were the driving forces for the extended DLVO theory. These differences were noted in biochemical processes and biological effects studies. The biological systems look like a huge complex colloidal dispersion system consisted of particles (cells) mostly of lipid composition (phospholipids, cholesterol, etc.) that are surrounded from a dispersion media with specific physicochemical properties depending on the biological compartment they are into (outer cellular fluid, intracellular fluid, blood plasma) and interact with each other, preserving stability homeostasis. For this reason, it is possible to explain various effects with the same physical laws that are used in colloids. DLVO theory has been proven to be inefficient to describe these interactions, since, according to various studies, the theory loses the prediction in salt concentrations higher than 5×10^{-2} M (organisms have high salt concentrations), for distances between particles less than 100 Å and for ionic dispersion media. Evolving nanotechnology/nanoscience, nanocolloidal dispersion systems have nanoparticles of less than 100 nm size. It is obvious that the transition in *nano* scale classification brings out the need of incorporating DLVO theory with various physical parameters that will contribute into the most adequate theoretical description of the experimental data. Also, DLVO theory cannot describe the particle behavior whose surface is not charged, as well as the particle behavior with ζ-potential when divalent or multivalent ions have been added into the system.

In both occasions, and contrary to DLVO theory that exclusively refers to the attractive *van der Waals* forces and the electrostatic repulsion forces, the existence of a different kind of interaction has been noted, the hydration interaction (Fig. 2.12).

When colloidal particles have a surface with certain hydrophobic grade, different experimental techniques and theoretical approaches have focused to the existence of hydration energy E_{hyd} that is in general a repulsive force. The hydration forces act in

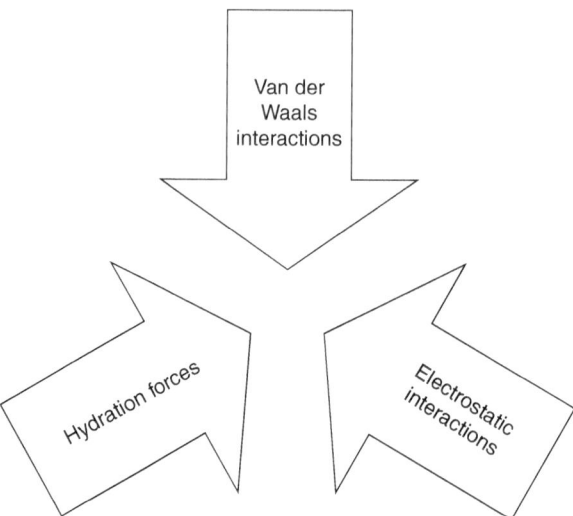

Fig. 2.12 Interaction between particles according to the extended DLVO theory

a distance of 1 nm. A hydration force is included in the interactions between particles and polyelectrolyte solution. This is happening due to the water molecule orientation onto the particle surface. The hydration forces are present in polar environment. Most systems in nature consist of water, and it is crucial to mention them and their contribution in order to describe the behavior of a colloidal dispersion system. It has been noted that the interaction energies can be divided into two different categories according to the nature of the two interacting membrane surfaces, the attractive and the repulsive. They are directly linked to the interface distance (distance between two particles) and related to the particle size and the width of the lipid or polymer or any other composition of the membrane.

The hydration forces are generally strong but short-range repulsive when the surface particle is highly hydrophilic. The hydration energy in this case is necessary for the water molecules to move from these hydrophilic surfaces. The force gradually weakens depending on the interfacial distance, taking into consideration a small depreciation constant of 1.7–2.5 Å.

In the case where the particle surface is hydrophobic, the hydration forces are attracting due to the water realignment onto the membrane surface. At this point, it must be mentioned that the hydrophobic properties of a surface with negative ζ-potential increase when the multivalent cations are added to the dispersion system and are attached to the particles surface. As every membrane surface consists of a mixture of smaller hydrophobic and hydrophilic surfaces, the term of total hydration interaction energy has been introduced. This is the total of the repulsive hydration energies and the attracting hydration energy. The total hydration potential is given from the following equation:

$$(Hyd = \text{hydration})$$

$$V_{Hyd} = V_{Hyd}^{R} + V_{Hyd}^{A} \qquad (2.8)$$

So the total potential V_T is given from the following equation:

$$V_T = V_A + V_R + V_{Hyd} \qquad (2.9)$$

Extended DLVO contains other forms of interactions, one of which is the hydration [1]. The hydration interactions appear in intermembrane distances of 20–30 Å and increase exponentially when the distances become less than 20 Å. Even though it was expected when particles of 10 Å distance, always compared to their size, merge and create aggregations, this seemed to be necessary but not sufficient, since the repellant hydration interactions at this point have great values. Therefore, it has been noted that these interactions predominate over the rest in distances less than 25 Å (mostly between 10 and 25 Å) and should be taken under consideration. In the special case where the particles consist of *zwitterionic* biomaterials (do not have a clear charge) and are dispersed in water, then the electrostatic forces do not act, and the attracting *van der Waals* interactions predominate. This means that aggregation and general phenomena of system thermodynamic collapse are inevitable.

Experimental data of various research groups have shown that, if large concentrations of ionic salts are added to this system, there is no thermodynamic collapse of the system. Especially, in the case where the surface of the system structural unit major contact, i.e., phospholipids, with water molecules is great and characterized of a great polarity, the hydration forces act in a greater distance (in comparison with 20 Å), more or less 25 Å. At this distance, *van der Waals* interactions do not predominate over the rest of the forces.

Extended DLVO is connected to its contribution to the hydration energy theoretical calculations. Hydration energy explains the thermodynamic stability or instability of a colloidal dispersion system. DLVO theory is the qualitative approach to colloidal dispersion system stability and is related to physical stability and the aggregation mechanisms.

In colloidal dispersed systems, the electrokinetic potential is referred to as ζ-potential and denoted using the Greek letter ζ (zeta) [4]. It gives an indication concerning the physical stability of a nanosystem. The ζ-potential of a nanocolloidal system should be ± 30 mV in order to stabilize the nanosystem based on electrostatic forces. By combining electrostatic and steric forces, the value of ζ-potential could be around ± 20 mV. It should be mentioned that ζ-potential is the electric potential at the "slipping plane" of the particle and not on its surface, which corresponds to the Stern potential (ψ_0) [13].

ζ-potential is not the particles' surface potential ψ_0, but these two are related (Fig. 2.13):

• ζ-potential can be used as a credible parameter that functions as a guide to quantify the repulsive forces between particles.

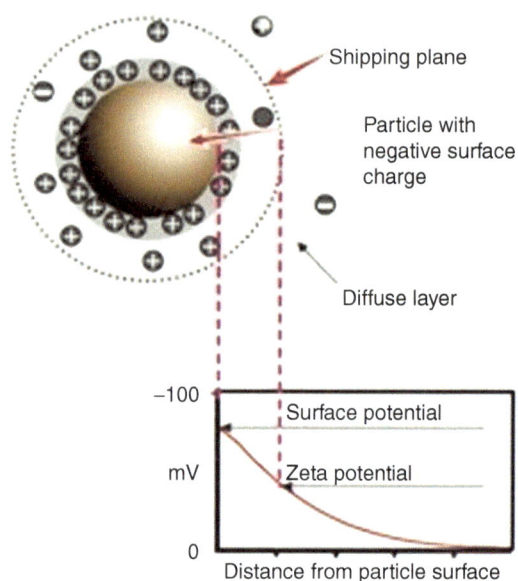

Fig. 2.13 The ζ-potential is the potential of slipping plane – it is not the surface charge (Adapted from [13])

- ζ-potential is considered as an indicator to predict the physical stability of a dispersed colloidal system.
- Henry's equation is used for the ζ-potential (ζ) calculation using the u_E:

$$u_E = \frac{\zeta \varepsilon}{4\pi n} f(\kappa\alpha)$$

(2.10)

The Debye length in a colloidal dispersed system is denoted with κ^{-1}:

$$\kappa^{-1} = \sqrt{\frac{\varepsilon_r \varepsilon_0 K_B T}{2N_A e^2 I}}$$

(2.11)

I is the ionic strength; ε_0 is the permittivity of free space; ε_r is the dielectric constant; K_B is the Boltzmann constant; T is the absolute temperature in K; N_A is the Avogadro number; and e is the elementary charge.

Where $f(\kappa\alpha)$ ranges between 1 (for low $\kappa\alpha$ values) and 1.5 (for high $\kappa\alpha$ values), ε is the dielectric constant of the continuous phase and n the viscosity.

In systems where $\kappa\alpha$ is low, the equation can have to following form:

$$u_E = \frac{\zeta \varepsilon}{4\pi n}$$

(2.12)

It is obvious that the physicochemical characteristics of particle surface in colloidal dispersion systems are critical parameters for the interaction determination with the plasma proteins. Measuring the particles' ζ-potential offers information for the total charge and how it is affected from the surrounded environment, e.g., electrolytes, pH, proteins, etc. Also, extremely valuable are the information for particles' ζ-potential when in a specific environment, with specific physicochemical characteristics, for both the enclosure and the release studies of the bioactive molecules- taking into consideration the bioactive molecule physicochemical characteristics in the environment where the particles are dispersed. The ζ-potential, as previously mentioned, is not the surface potential ψ_0 (Stern's layer) of the particle, but is related to it. The ζ-potential changes by surfactants or other substances added and affects colloidal stability. In the literature, the stability studies are described by addition of various substances, e.g., addition of phosphates in a dispersion with positive ζ-potential particles can change (reduce) the particles' ζ-potential. DLVO theory is only for electrostatic repulsion, while colloids can be stabilized with repelling forces that are a result of macromolecules and surfactant adsorption onto their surfaces (stereochemical or steric stabilization). In water media, the adsorbed molecules will be hydrated; therefore, the repulsion between hydrated surfaces (repulsion between hydrated surface steric stabilization) is a result of the entropic effect or osmotic effect or enthalpic stabilization. We have to note that ζ-potential is not only an indicator of the stability of the nanocolloidal system, but it contributes to the biophysical behavior of nanoparticles which could be used as carriers of a bioactive molecule for therapeutical purposes. Additionally, regulatory considerations aim to provide stability proofs of the final drug formulation's physical characteristics, and measurements of the ζ-potential should be included into the dossier submitted for evaluation from the authorities.

The entropic effect causes, to chains' adsorbed molecules, the loss of free movement. The approach of the two particles with the adsorbed stabilized chains leads to steric interaction when the chains interact. The steric effect appears only when $x = 2r$ when the interaction increases with the distance reduction. The reduction of conformational freedom leads to a negative entropy change $(-\Delta S)$. Each chain loses its conformational freedom, and therefore, free energy system increases, leading to repulsion. The steric effect depends on:

- The chain length of the adsorbed molecule
- The chain-solvent interactions
- The chain number

The osmotic effect arises as the macromolecular chains of the neighbor particles increasing their concentration in the overlapping area. The resulting repulsion is due to the solvent osmotic pressure as the solvent tries to dilute the saturated area. This can only be achieved when particles are drawn away.

According to particle approach, the hydrated water molecules that are adsorbed are being detached. This fact is the cause of the increased enthalpy leading to repulsion. More examples for the systems stability in Pharmaceutical Nanotechnology (e.g., liposomes) using DLVO theory will be mentioned in Chap. 5.

2.3   Basic Principles and Physicochemical Characterization of Nanocolloidal Dispersion Systems

2.3.1   Microscopy

2.3.1.1   Optical Microscopy

Optical microscopy is a widely known and commonly used technique for the direct observation of nanosystem shape transformation in real time, for example, liposomes. Microscopy of high-intensity dark-distant field is used for nanosystem imaging, for micrometer-sized systems observation, and for characterizing the lipid membrane interactions, i.e., liposomes, surfactants, etc. Nanosystem "optical trapping" is a useful technique for their manipulation for studying their properties and their spectroscopic observation. Optical forces create submicron-scale deformations in small particles measuring their elasticity and rigidity.

During the past decades, multiple technical approaches have been made for optical analysis in nanometer scale. Using confocal fluorescence microscopy and 3D confocal microscopy, scientists have achieved substantial improvement in analyzing nanosystems.

2.3.1.2   Scanning Probe Microscopy

Scanning probe microscopy (SPM) is an effective analytical technique that allows the visualization of proteins, nucleic acids with lipid membranes, and nanoparticulate systems like liposomes and of their complexes with biomolecules. Scientists can also measure the width of lipidic vesicles, and by using the tunnel microscopy, they can define their shape and diameter. The advantages are high analysis and functionality in air and liquid vacuum. The samples cannot be damaged due to presence of high vacuum and because they are not being subjected to specific treatments. There are developments and improvements in the way the probe contacts the sample and probe structural and/ or chemical modifications that aim the friction reduction between the probe and the sample. Atomic force microscope (AFM) is the most common scanning probe microscopy for liposome and lipidic vehicle studies encapsulating bioactive molecules.

International literature refers to various analytical techniques that have been used to characterize nanoparticles. Among them, the widely used are chemical, thermodynamic, hydrodynamic, microscopic (microscopy), spectroscopic, diffraction, and light scattering techniques and, finally, chromatographic techniques.

Scientists have used these techniques in order to physicochemically characterize nanosystems and nanoparticles like liposomes, dendrimers, polymers, etc. More in particular, this characterization refers to the study of nanoparticle properties, i.e., measurement of the width of the liposomal lipid membrane, the ability of allowing biomolecules to penetrate their membrane, the bioactive molecule loading ability for therapeutic purposes, but also for tissue diagnosis and imaging. Also, the nanosystem stability in biological fluids over time is a part of studying and characterizing these properties according to the methods mentioned before [12].

2.3.1.3 Electronic Microscopy

Electronic microscopy (EM) is used to study nanoparticle shape and its changes during time or to investigate the changes of the physicochemical characteristics of nanoparticles. Transmission electron microscopy (TEM), scanning electron microscopy (SEM), reflection electron microscopy (REM), and scanning transmission electron microscopy (STEM) are techniques providing information for vesicles, e.g., lipidic carriers in terms of shape, bilayer thickness, and distance between layers in nanosystems. They can also offer information concerning nanosystem aggregation process. The restrictions of the electronic microscopy methods are related to the need of high-vacuum conditions in order to handle or to edit the samples (i.e., gold or carbon cover) that might cause non-reversible changes or sample damage [12].

2.3.1.4 Atomic Force Microscopy (AFM)

AFM belongs to the scanning probe microscopy family, therefore, is almost an invasive technique, as the probe forces are applied across the nanoparticulate systems surface and might cause deformations. It uses silicon-based needles of atomic sharpness and applied to study the surfaces' topography at atomic level [17]. On the contrary, the "optical trapping" in order to measure the elastic properties does not acquire nanoparticulate systems adsorption in any substrate, therefore, without risking deformations [22]. The optical probe's contact surface is a lot smaller than liposome surface, while the optical force analyzing depth, much larger. So, their combination of these two methods is considered the ideal one for the optimal study of nanoparticulate shape transitions. AMF is a technique that can study the shape, the mechanical deformations, and the micro-engineer properties of liposomes, in nanometer scale. AFM is an important topographic and spectroscopic tool as spectroscopy of force deformation can measure the bond forces between biomolecules and electrostatic forces, offering information for nanoparticulate systems' stability, their elastic properties, or relevance forces, without causing structural disorder. AFM is used for detection of viruses for early diagnosis of vital infections (Bioforce Virichip (Ames, IA, USA) and offers challenges for developing innovative sensors for diagnostic purposes. It can be concluded that AFM is a probe microscopic technique for applications at nanoscale analysis [24].

Electronic microscopy techniques also include the following:

Magnetic resonance force microscopy
Near-field scanning optical microscopy
Characterizing nanostructures with Halo™ LN100 technology
4pi microscope [12]

2.3.2   Thermal Analysis

Thermal transitions appearing during a nanosystem life and especially to nanosystems to be used in the field of health are extremely important. Their thermotropic behavior is related to nanoparticle dispersion system thermodynamic behavior and is affected from temperature changes. It is obvious that the methods with which scientists can study these changes contribute to stability understanding and nanoparticle rational design. According to International Confederation for Thermal Analysis and Calorimetry (ICTAC), thermal analysis can be defined as: "the total of analytical techniques where scientists can measure a property of the sample or its products in correlation to temperature, while the sample is submitted in a programmed thermal process, in a designated environment."

Thermal analysis advantages in comparison to other analytical techniques are summarized in the following points:

- The sample properties can study in a wide temperature range, using various heating and cooling programs.
- The sample can be in any form: liquid, solid (amorphous or crystalline), or gel, using various crucibles/ sample holders made from different materials (e.g., platinum, aluminum, etc.).
- Smaller sample quantity is required, i.e., 0.1–10 μg nonhomogenous samples in combination with the small quantities required for analysis can cause problems related to the sample representativeness.
- Sample preparation is simple and refers to grinding and homogeneity of solids.
- The atmosphere around the sample is defined from the analyst.
- The required time of analysis is ranged between few minutes to few hours.
- The acquisition and function cost of thermal analysis tools is small.

The disadvantages of thermal analysis in comparison to other analytical techniques (i.e., spectroscopic or electrochemical) are summarized in the following points:

- They are nonselective techniques. In many occasions, extra analysis is needed, i.e., IR (infrared) spectroscopy or microscopy analysis. Also, they can be combined with additional techniques, for example, the evolved gas analysis (EGA). Featured examples include the combination of mass spectrometry (MS) and Fourier transform infrared resonance, thermogravimetry-MS, and TG-FTIR, accordingly.
- Sensitivity and accuracy in quantitative measurements of thermal analysis techniques do not exceed ±2 %.

Thermal analysis techniques can be classified according to the property under investigation. The most common used are the following:

- Thermogravimetry(TG)/differential TG (DTG)
- Differential thermal analysis (DTA)
- Differential scanning calorimetry (DSC)
- Thermomechanical analysis (TMA)

- Dynamic mechanical analysis (DMA)
- Evolved gas analysis (EGA)

Many applications of these methods refer to analysis of polymers, minerals, metals, alloys, ceramics, glass, building materials (i.e., concrete), catalyst agents, explosives, fireworks, lipid oil and soaps, lubricants, fuels, medicines, food, fabrics, fibers, biological samples (e.g., bladder stones), flame retardants, etc. [10].

2.3.2.1   Differential Scanning Calorimetry

The test sample and the reference sample that may – or may not – acquire thermal phenomena in the designated temperature range are heated at the same time with the same rate. The temperature rises exponentially with time and the difference between their temperature grades equals zero. The study sample is subjected to thermal phenomena; the control system sensitizes and offers greater or lower temperature to the study sample, so the temperature would be the same with the one of the reference sample. The calorimeter records the differential heat as a temperature function. The energy change that is offered to the system in the form of heat during time (dq/dt) is proportional to the heat capacity of the study sample.

When a thermal phenomenon occurs, the differential heat that is recorded has the shape of a curve, where the concave face downward or upward, depending if the phenomenon is endotherm or exotherm, respectively (the opposite might occur as well depending on the device or analytical program that is being used). The recorded curves are the thermogram of the study sample. The transition enthalpy (ΔH) – from one state to another – is the amount of energy transferred as heat under stable pressure conditions and is related to the system's heat capacity (C_p). It can be calculated as the curve integral that results from the differential heat as a temperature function and is recorded from the calorimeter:

$$\Delta H = C_p dT \tag{2.13}$$

Apart from the change of transition enthalpy (ΔH), the differential scanning calorimeter allows the determination of important thermodynamic parameters. The change in Gibbs free energy of a process corresponds to the equation:

$$\Delta G = \Delta H - T\Delta S \tag{2.14}$$

where ΔH is the enthalpy change and ΔS is the entropy change accordingly during the process. When the change of Gibbs free energy (ΔG) equals zero, the system is a mixture of two equivalent states. The temperature where the two states are equal is temperature T_m. At T_m temperature, if the change of Gibbs free energy (ΔG) between two states equals zero, then:

$$T_m = \Delta H / \Delta S \tag{2.15}$$

The main transition temperature T_m is the temperature where the system has the maximum heat capacity C_p.

During phase transitions that are described with symmetrical graphic curves, the corresponding main transition temperature T_m represents the temperature where the phenomenon is half completed. However, in asymmetrical curves, the temperature T_m does not correspond to the middle of the phase transition curve and, instead of T_m, is used $\Delta T_{1/2}$ that corresponds to the middle of the asymmetrical phase transition curve range and represents the actual temperature where the phenomenon is half completed.

The shape of the differential temperature curve versus temperature offers important information related to the occurring phase transition. The temperature range that correlates to the middle of the curve height is related to biomaterials or material purity, quality, and cooperativity in the system during transition phase or phases versus temperature changes. The presence of biomaterials that function as additives (non-thermodynamic cooperative biomaterials) results in the expansion of phase transition curve (i.e., asymmetry, multiplicity). The result indicates the kind of the interaction the additive has with the studying nanosystem. Usually, the phase transition, i.e., in a phospholipid bilayer – in case of liposomes – contains intermediate phase sequences, until it switches from one phase to another that is thermodynamically stable, for a specific time, in a specific temperature change range.

These intermediate phases, the "mesophases," occur due to the formation of areas within the nanosystem structural units' different orientation, i.e., phospholipids in case of liposomes due to *flip-flop* phenomena, *lateral motion*, and diffusion into the double layer. These intermediate phases can be created due to additional infections, materials, or biomaterials that do not have the demanded cooperativity [10] with the actual structural biomaterials and create thermodynamically non-favorable mesophases. This pattern can be correlated, as previously mentioned, with the presence of additional biomaterials or biomolecules from the environment that can develop – in a cellular level – unwanted *mesophases* onto the cellular membrane, retaining its functionality and leading to diseases according to biophysical approaches.

In case of liposomes, the phospholipids' numbers that exist in each phase as well as the dispersion of phases that coexist in the phospholipid bilayer define the range of the total transition that is expressed as cooperativity between biomaterials. The cooperativity of biomaterials during transition phases is usually expressed as the temperature change that corresponds to the one half of the curve, known as $\Delta T_{1/2}$. The onset of the thermal phenomenon is the temperature, where the extension of the straight baseline intersects with the extension of the straight line from the top of the curve that corresponds to T_m to the start of the thermal transition, and it is characterized as T_{onset}, while at the end of the phenomenon it is characterized as T_{endset}, accordingly [5]. It is worthy to note that lipidic nanoparticulate systems such as liposomes have thermodynamic advantages over other nanosystems such as polymeric nanostructures. Their thermodynamic profile does not meet to the thermodynamic equilibrium, but they behave as a kinetically trapped nanosystem, contrary to those nanosystems (e.g., polymeric nanosystems or microemulsions) that they are affected when changes in their environment occurred. However, lipidic nanosystems preserve their physicochemical properties and could be attractive drug delivery nanosystems from a physicochemically and thermodynamically point of view [15].

DSC measures the change in the heat flow that is emitted or absorbed between the sample and the reference substance, when they are both heated in a controlled rate. It is interesting to mention that DSC measures the change of a property than the change of temperature rate that is usually emitted due to the increased temperature of the sample. When there is no change in the temperature of the sample, the change on the heat flow cannot be measured, except if a chemical reaction takes place.

There are two DSC types:

- Heat-flux DSC where the sample and the reference substance receptors are heated in a controlled rate in the same space. The appropriate connection of the thermal balances (sensors) between the sample and the reference substance allows the measurement of the heat flow that is independent of the specimen properties. The sample's enthalpy change is defined after the comparison of the calibration of a known substance and the one of the sample.
- Power-compensation DSC, where the two receptors of the sample and the reference substance are heated from two different sources (heat resistance) in a stable temperature. The additional heat that is needed to be added or eliminated during the transition to the one or other receptor to achieve the same temperature balance offers the transition energy thermodynamic curve.

The greatest advantage of the heat-flux DSC over the power-compensation DSC is that the first type can function in higher temperature, while the second one has greater sensibility due to direct energy measurement onto the sample and the reference substance. The measurement values that are obtained with differential scanning calorimetry allow the determination of thermal capacity, transition temperature, glass transition temperature, movement data, etc.

2.3.2.2   Thermal Behavior and Calorimetric Characterization of Materials and Biomaterials

Materials or biomaterials can undergo the following thermal changes: melting, crystallization, polymorphic changes, boiling, sublimation, dehydration, solid-solid transition, and glass transition (amorphous and/or semicrystalline materials). These changes can be endothermic or exothermic (Table 2.5). It is important for all DSC tests to state the onset temperature (T_{onset}), the extrapolation of T_{onset} (T_{exp}), the endset temperature (T_{endset}), as well as the scanning rate.

Crystalline materials undergo a first-order melting transition from the organized solid phase to the unorganized liquid phase. DSC offers precise melting temperatures that might be impossible to achieve with traditional methods. For example, in the case of carbamazepine, two different transitions are described in DSC that are related to the melting temperature of a polymorph that recrystallizes in a second more stable crystalline form and has a second melting temperature. Using the classic method, only the second melting temperature is observed. Melting is an endothermic transition

Table 2.5 Enthalpy changes/ transitions during the differential scanning calorimetry (DSC)

Endothermal	Exothermal
Glass transition	Decomposition
Melting	Crystallization
Evaporation	Condensation
Disruption	Disruption
De-solvation	Solvation
Decomposition	Oxidation

where the sample collects energy. For small or 100% crystal molecules, the curve top is sharp and the melting temperature is the temperature at the top of the curve. For larger molecules, like polymers, the melting temperature is wider since the curve top is broader and proportional to the polydispersity of the molecule. Often, melting temperature is the onset or endset temperature of the phenomenon, with the endset temperature being the most rational one as it defines the temperature where all crystals have been melted. Conventionally thought, T_m is the temperature at the top of the curve. The area under the curve represents the total of temperature absorbed during the melting process. The area under each first transition in a DSC diagram offers the total eliminated/absorbed temperature during transition. DSC data measures thermal capacity or temperature change in time dq/dt. Integrating, we receive total temperature q or otherwise enthalpy ΔH:

$$\int_{T_2}^{T_1} \left(\frac{dq}{dT} \right) dT = q = \Delta H \tag{2.16}$$

The endothermic melting curve can be described with the following parameters:

- Onset temperature (T_{onset}) is the temperature where the phase transition starts depending on the temperature change.
- Main temperature is the temperature at the top of the curve.
- Endset temperature (T_{endset}) is the temperature where phase transition is completed and the curve meets the baseline.
- Extrapolated onset temperature is the temperature where the curve extension of the endothermic or exothermic top meets the baseline.

The purity of crystal substance can be calculated from the enthalpy and the melting temperature using the *Van't Hoff* equation:

$$T_S = T_e - RT_e^2 x / \Delta H_0 F_i \tag{2.17}$$

where T_s is the sample temperature in equilibrium (K), T_e is the melting temperature of the pure substance (K), R is the gas constant (9.314 J/mol/K), X is the molecular impurity fraction, ΔH_0 is the enthalpy of the pure substance (J/mol), and F is the sample fraction in the melting temperature T_s. When T_s is depicted as the function of $1/F_i$, linearity is expected. Despite all these, *Van't Hoff* diagrams have a curve shape due to thermal delay and formation of solid solutions. Also, an important percentage of the substance can be melted before a measurable energy flow is observed in the DSC.

To limit these kinds of phenomena, a constant is usually added to the measured areas for *Van't Hoff* diagram correction. The curve section being used in the calculations can affect the purity value being received. To limit the error from the thermal delay phenomena that take place during the onset of melting transition, it is recommended to use various fractions, 3–30%, 3–40%, and 3–50% during purity calculations. This way it can be concluded whether there is a slow decomposition and decay. Generally, the measurements for purity are made for a variety of fractions around 10–50%. The requirements for defining purity are low scanning speeds (<5 °C/min, preferably 2 °C/min) that limit the thermal delay phenomena and small samples (1–3 mg). To avoid solid-solid transition states, like polymorphic changes, the test of various scanning speeds is recommended. If the purity values are within experimental error limits, then the substance decay or the transition in other polymorphic states is not taken under consideration. Generally, accurate results come from samples with purity levels between 98 and 99.5%. The advantage of this method is the measurement speed. Small amount of sample is needed and the impurity type doesn't need to be known. The purity determination can aid into the evaluation of pharmaceutical batches' quality and has been used for the determination of the amorphous content and the organic substance degradation.

The understanding of crystalline and amorphous nature of bioactive molecules and biomaterial excipients is very important for the development of pharmacotechnological formulations. In case of crystalline molecules, a percentage can change into amorphous during the transition process, for example, during liquid granulation and sieving. This can affect the final product, especially its physical and chemical stability. Using DSC, crystallization looks like an exothermal transition, and samples that crystallize during heat contain automatically an amorphous percentage. Despite all these, there are amorphous ingredients that do not crystallize. The energy released during crystallization relates to the energy formation of the molecule's crystalline lattice. This way, DSC can be used to quantify the crystallization of lyophilized or sieved substances.

Polymorphism is defined as the phenomenon where solid compounds of the same molecular structure are crystallized in more than one forms. As a result, the polymorphs have different melting points, viscosity, solubility, physical and chemical stability, different physical properties, and bioavailability. The choice of a known stability polymorph is of great importance for the development of stable pharmaceutical products. Polymorphs are divided into categories: enantiotropic and monotropic. Enantiotropic polymorphs produce a single crystal form in a huge range of pressure and temperature grades. The stable form has a greater melting point and enthalpy than the less stable crystal in all different grades of pressure and temperature. If the form with the greater melting point has the smallest enthalpy, then the polymorph is characterized as enantiotropic [1].

DSC can be used as a quick method for stability test, in combination with chromatographic techniques. The advantages of DSC include the measurement speed and the small amount of substance that is needed. The decomposition kinetics can be defined by the use of Arrhenius equation and DSC experiments. During isothermal decomposition experiments, the sample reception (pan) temperature is retained stable in various temperatures for various time intervals. Then, the sample is scanned in a high speed to avoid additional decomposition.

2.3.3 Light Scattering Techniques

2.3.3.1 Measurements of Size Distribution and ζ-Potential of Nanocolloidal Dispersion Systems

The evaluation of measurements related to size distribution and ζ-potential of particles in a nanocolloidal dispersion system is important as they define the stability of the pharmacotechnological product and its effectiveness as a pharmaceutical product. Light scattering is the method used. This method is used for the evaluation of nanoparticles' parameters, i.e., diffusion coefficient, size, and molecular mass. The evaluation of particle size with light scattering technique is based on the following physical principles:

• Light diffraction. The principle that light scattering is based on when it falls onto small-size particles is related to the intensity of the diffraction radiation that depends on the diffraction angle, the particle optical properties, and their size. The particle size is calculated according to the distribution of the diffraction radiation using computer programming.
• Single particle scattering. Single particle scattering is based on the principle that the particle size is evaluated basically from the intensity of the scattering radiation and secondly from the diffraction angle.

The equipment angle is stable, and according to particle size distribution and particle number, the median diameter and the size dispersion are calculated.

Literature has shown that the therapeutic value of a nanoparticulate system with a loaded bioactive molecule is related to the bioactive molecule release kinetics from the nanoparticle colloidal system core. It is obvious that the core composition and physical characteristics affect the system's physicochemical characteristics and, therefore, the bioactive molecule release. The nanoparticle physicochemical characterization is a process that includes high technological studies and requires great expertise in result evaluation. These techniques relate to measure nanoparticles' size and size distribution, to ζ-potential measure, to shape studies, to bioactive molecule loading capacity, to physicochemical characterization of a bioactive molecule nanosystem, etc.

It must be mentioned that the parameter study of nanosystem characterization should be achieved when nanoparticles are empty from the bioactive molecule. These parameters should be compared to the ones attained when the bioactive molecule is loaded to the nanosystem. The nanoparticles' size – which is defined according to the diameter of a sphere which has equivalent diffusion coefficient (D) to that of the nanoparticles – and size distribution are important parameters that affect the system's therapeutic result. The greater surface for the same volume or mass is an advantage that nanoparticles of a nanocolloidal dispersion system have over microparticles.

The sphere volume is given from the following equation:

$$V = \frac{4}{3} \pi R^3 \qquad (2.18)$$

The sphere surface is given from the following equation:

$$S = 4\pi R^2 \qquad (2.19)$$

The relation between surface radii of a particle is given from the following equation:

$$S = \frac{3V}{R} \qquad (2.20)$$

The equation above shows that the smaller the nanoparticle size (diameter D or radius R) of a colloidal dispersion system, the greater the overall surface. This fact facilitates the bioactive molecule release from nanoparticles. The nanoparticle size and size distribution facilitate their passage through barriers and help in the nanoparticle interaction with the cells and their fusion, bioadhesion, or absorption to the cell membrane. The method used is the laser light scattering. The polydispersity index (PDI) is a measurement of nanoparticle uniformity. The study of nanoparticle shape according to their size cannot be completed with the use of optical microscopy, as they are hardly distinguished.

Scanning electron microscopy (SEM), atomic force microscopy (AFM), and fluorescent microscopy are used in case of colored particles, i.e., rhodamine. Nanoparticle's ζ-potential is also a very important parameter that should be studied. ζ-potential can affect the system's stability and its interaction with cells whose charge is most of the times negative. Therefore, nanoparticles can be designed having the appropriate ζ-potential that will affect the load or the release of bioactive molecules from the nanosystem. Depending on the total charge of the bioactive molecule, scientists can design its interactions with nanoparticles and develop an appropriate carrier for controlled and targeted release.

The question is whether the bioactive molecule's physicochemical characteristics when it is loaded into the nanosystem and then released from it are the same as before its entrapment. This question is crucial and the response to it assures the effectiveness of the final medicine. It is known that the molecule's chemical characteristics can change according to the environment that they are in. This is referred mostly during the configuration study after their interaction with the nanoparticle components into the nanosystem. The conformation properties of a bioactive molecule or a biomolecule (proteins, peptides, etc.) are crucial factors especially for when it reaches the designated target.

Also the presence of enantiomers and the possibly different release from the polymer matrix or their simultaneous release can affect the pharmacological result. The release rate of each enantiomer offers advantages or disadvantages to the total therapeutic result. Studies referring to techniques like DSC can offer information for the interaction of molecules with polymer matrices. Bioactive molecule conformational properties in lipidic drug delivery systems and their release affect probably not the bioactive molecule but definitely the nanosystem thermotropic properties, and this affects the total system behavior during release process.

The particles that are dispersed in a liquid are in constant move (Brownian motion). The motion speed depends on their size, the temperature, and the viscosity of the medium. The size distribution is proportional to their stability. When the particle size distribution is stable in time, it means that the dispersion system is stable as well. In case where size distribution increases with time, the nanosystem can be characterized as unstable and probably inefficient as pharmacotechnological formulation for loading bioactive molecules. Using dynamic light scattering technique (DLS), the measurement of this movement is possible through the data collected for the scattered light. By measuring Brownian motion, the diffusion coefficient (D) and the particle size can be calculated. The diffusion coefficient (D) and the particle size as it is expressed from the average hydrodynamic diameter are mathematically related with the Stokes-Einstein equation:

$$D = \frac{K_B T}{6\pi\eta R_h} \tag{2.21}$$

where T is the absolute temperature, K_B is the Boltzmann constant, η is the medium viscosity, and R_h is the particle hydrodynamic radius. DLS measures the particle size that is related to the particle movement in a liquid, therefore, Brownian motion. The result is the diameter of an equivalent sphere with the same diffusion coefficient (D) with the particles that exist in the sample. The diameter of the equivalent sphere is called hydrodynamic diameter. The diameter of a tube-shaped particle with dimensions 100 and 20 μm is defined according to diffusion coefficient of an equivalent sphere with a dimension of 39 μm. Therefore the diameter of a tube-shaped particle according to the determination of the diffusion coefficient of the equivalent sphere is 39 μm.

The hydrodynamic diameter is a little larger than the real diameter of the sample particles due to solvation and the interactions between particles.

ζ-potential is nanoparticle stability measure (Table 2.6) and its measurement contributes to their physicochemical characterization.

ζ-potential is measured through electrophoresis. The sample is placed in the electrophoresis cell; an already known electric field is applied and lights from crossed laser beams. Particles moving to the lighted sample mass scatter the light.

Table 2.6 Approximate values of stability (ζ-potential) of nanoparticles

	ζ-potential (mV)
Maximum agglomeration and precipitation	0–3
Strong agglomeration and precipitation	5
Threshold of agglomeration	10–15
Threshold of delicate dispersion/suspension	15–30
Moderate stability	30–40
Fairly good stability	40–60
Very good stability	60–80
Extremely good stability	80–100

Adapted from Koutsoulas et al. [13] with modifications

ζ-potential relates with electrophoresis movement and is expressed through Henry's equation:

$$\zeta = \frac{4\pi n u_E}{\varepsilon f\left(\kappa\alpha\right)} \tag{2.22}$$

Where u_E is the electrophoresis movement, ζ is the ζ–potential, ε is the dielectric constant, n is the viscosity, $f(\kappa\alpha)$ is the Henry's function, α is the radius of the particle, and $\kappa\alpha$ is the fraction of particle radius to bilayer width. For particles in polar solvents, the maximum value of $f(\kappa\alpha)$ is 1.5 (Smoluchowski approach), while for particles in nonpolar solvents, the minimum value of $f(\kappa\alpha)$ is 1 (Hückel approach).

2.4   Freeze Drying of Nanocolloidal Dispersion Systems

The target of pharmaceutical technology science is to find the appropriate pharmacotechnological formulation, in order for the final product to maintain its effectiveness, its safety, and its commercial appearance. Freeze drying is the method that assures the physical and chemical stability of the pharmaceutical products especially pharmaceutical dispersions and solutions. It is the mildest method where drying is achieved with water sublimation. The product is cooled down in temperatures between −20 and −60 °C, so the water from the liquid form turns into solid form. Following, the ice sublimates in a special device that functions under vacuum. It turns into air form (water vapor) without passing by from the liquid form. This method is expensive, demands complicated devices, and is not being used in routine processes. The freeze-drying process started in the 1930s and was industrially developed during World War II due to the need of stable pharmaceutical products.

Production of freeze-dried products started due to the following needs:

- Stability of pharmaceutical products.
- Receiving of a solid (amorphous) from a water solution or dispersion system when its crystallization is not possible.

- Sensitive in temperature or in solutions products.
- Use of mild temperature and pressure conditions. Avoid breakdown of products.

Freeze drying is:

- The transformation of liquid product in a solid product through the process of sublimation.
- The water of dispersion or solution of the product is removed from the ice with sublimation.
- The solid residue includes bioactive molecules and biomaterial excipients and a very small amount of water (0.5–2.0%).

The freeze-drying procedure design is a subject of innovative research and can be patented. The temperature and the pressure that correspond to the triple point relate to pure water. The conditions for the products to be lyophilized are unknown for the original product. The procedures that assure the sublimation process and limit the dispersion process of the products should be investigated. The experimental design of the freeze-drying process in pharmaceutical industry is related to the kind of products to be lyophilized, their sensitivity, and the duration of their stability that the pharmaceutical industry wants to achieve.

The freeze-drying process of nanocolloidal dispersion systems is described as follows:

- Solution or colloidal dispersion production
- Filling of the containers, usually vials
- Plastic cap placement onto the vial orifice (partial closure)
- Vial placement in freeze dryer under aseptic conditions
- Freeze in previously stated critical temperature for each product
- Use of appropriate temperature and pressure to sublimate the ice from the final product
- Application of appropriate temperature and pressure discarding the bound water from the product
- Automated vial sealing
- Vial discarding under aseptic conditions and aluminum seals placement onto the orifice

Characteristics of a freeze-dried pharmaceutical product:

- Lyophilic cake
- Satisfactory mechanical resistance
- Uniform color
- Satisfactory drying
- Satisfactory porous
- Sterilized
- Pyrogenic-free
- Chemical and physical stability of the dry form after reconstitution

An important advantage of this method is that the final dry product must have the same volume with the solution prior to freeze drying. This will result in the ease of its reconstitution. This happens because of the existence of vacuum spaces during the sublimating of the substance's particles.

Products that are usually freeze dried belong to the following groups:

* Small or large bioactive molecule
* Bacteria
* Viruses
* Vaccines
* Plasma
* Fruits
* Coffee
* Flakes
* Products that are easily disintegrated in water

Freeze-drying process is applied only in a small number of pharmaceutical industries around the world as it requires highly controlled manufacturing conditions. Defining these parameters during the freeze-drying process demands expertise that is not widely known and high cost equipment, making the initial investment rather large. Freeze drying outmatches other drying techniques because the manipulation takes place in mild temperature and pressure conditions and, therefore, offers the following advantages:

* High-quality products that would have been impossible to be produced with any other method
* Limitation of product disruption, a fact that makes freeze-drying process the most suitable one for highly cost raw materials
* Easy and quick reconstitution of the final product after the addition of the appropriate solvent as the dry lyophilized product has the same volume with the solution that was produced from
* Greater life expectance of the final product, as the dry product has greater stability than a corresponding solution

2.5   Summary

Biomolecules and biomaterials are the building blocks of nanosystems for pharmaceutical applications that meet the basic requirements of biodegradability and biocompatibility to be approved to the market.

Medicine is the commodity that is composed of the bioactive molecule and the pharmacological inert excipient. The medicine in its parts (bioactive molecule and excipients) and as a final marketed product should be considered as a *biomaterial*.

Biophysics is the science that deals with biomolecules, cells, and tissues. The biophysics and thermodynamics of cell membrane provide projection of the behavior of a lipid-based nanosystem.

The synergy regarding the biophysical behavior of artificial biomembranes and of cell biology has promoted nanoparticulate systems as leading nanosystems to deliver drugs to the target tissues.

The thermotropic behavior of the liquid crystalline state of nanosystems provides information which can be correlated with their functionality.

The phase transitions of the nanosystems' membrane due to the *flip-flop* and *lateral diffusion* behavior of its building blocks from which it is composed could be correlated with the *biophysical disease factor* as well as with the stability performance of the nanosystem.

Liquid crystalline state has properties affecting the visual, electrical, and magnetic behavior of a nanosystem.

The stability studies of nanocolloidal dispersion systems were measured based on DLVO theory and measurements of their charge (ζ-potential).

DLVO theory can be divided in its classical and extended mode. The first one deals with the repulsive and attractive forces, while the second one accommodates the hydration energy.

Thermal analysis techniques referred to the thermotropic behavior of nanosystems measuring the changes of the thermodynamic parameter. These changes reflect to their biophysical behavior.

Lipidic nanosystems such as liposomes are considered as kinetically trapped systems and are not at the thermodynamic equilibrium. Therefore they preserve their physicochemical profile.

The freeze-drying process of nanocolloidal systems is an approach to lyophilize and protect fragile and sensitive to the environmental conditions nanosystems, incorporating drugs.

References

1. Attwood D, Florence (2012) Physical pharmacy PhP. Royal Pharmaceutical Society, Oxford
2. Bangham AD, Standish MM, Watkins JC (1965) Diffusion of univalent ions across the lamellae of swollen phospholipids. J Mol Biol 13:238–252
3. Chapman D (1975) Phase transitions and fluidity characteristics of lipids and membranes. Quart Rev Biophys 8:185–235
4. Delgadro AV, Conzalez-Caballero F, Hunter RH et al (2007) Measurements and interactions of electrokinetic phenomena. J Control Interface Sci 309:194–224
5. Demetzos C (2008) Differential Scanning Calorimetry (DSC): a tool to study the thermal behavior of lipid bilayers and liposomal stability. J Liposomes Res 18:159–173
6. Derjaguin BV, Landau LD (1941) Theory of the stability of strongly charged lyophobic sols and of adhesion of strongly charged particles in solution of electrolytes. Acta Physicochim URRS 14:633–662
7. Evans WH (1980) A biochemical dissection of the functional polarity of the plasma membrane of hepatocyte. Biochim Biophys Acta 604:27–64

8. Evans WH (1979) Preparation and characterization of mammalian plasma membranes. In: Work TS, Work E (eds) Laboratory techniques in biochemistry and molecular biology part I. North-Holland, Amsterdam
9. Florence AT, Attwood D (1988) Physicochemical principles of pharmacy. Macmillan, London
10. Heimburg T (2007) Thermal biophysics of membranes. Wiley-VCH, Weinheim
11. Heurtault B, Saulnier P, Pech B, Proust JE et al (2003) Physico-chemical stability of colloidal lipid particles. Biomaterials 24:4283–4300
12. Kewal KJ (2008) The handbook of nanomedicine. Humana Press, Basel, pp 14–21
13. Koutsoulas C, Pippa N, Demetzos C et al (2012) The role of ζ-potential on the stability of nanocolloidal systems. Pharmakeftiki 24:106–111
14. Lasic DD (1993) Liposomes: from physics to applications. Elsevier Publishing Company, Amsterdam
15. Lasic DD, Papahadjopoulos D (eds) (1998) Medical applications of liposomes. Elsevier, Amsterdam
16. Lee AG (1977) Liquid phase transitions and phase diagrams: I lipid phase transitions. Biochim Biophys Acta 472:237–281
17. Mazzola L (2003) Commercializing nanotechnology. Nat Biotechnol 21:1137–1142
18. Mishra B, Bhavesh B, Tiwari S (2010) Colloidal nanocarriers: a review on formulation technology, types and applications towards targeted drug delivery. Nanomedicine 6:9–24
19. Nodre W (2003) Colloids and interfaces in life science. Marcel Dekker, New York
20. Okhi S, Ohsihima H (1999) Interaction and aggregation of lipid vesicles (DLVO theory versus modified DLVO theory). Colloids Surf B: Biointerfaces 14:27–45
21. Papahadjopoulos D, Bangham AD (1966) Biophysical properties of phospholipids II permeability of phosphatidylserine liquid crystals to univalent ions. Biochim Biophys Acta 126:185–188
22. Spyratou E, Mourelatou E, Makropoulou M et al (2009) Atomic force microscopy: a tool to study the structure, dynamics and stability of liposomal drug delivery systems. Exp Opin Drug Deliv 6(3):305–317
23. Verwey EJB, Overbeek JTG (1948) Theory of the stability of lyophobic colloids. Elsevier, Amsterdam
24. Wagner V, Dullaart A, Bock AK, Zweck A (2006) The emerging nanomedicine landscape. Nat Biotechnol 24:1211–1217

Part II
Nanotechnology in Imaging, Diagnostics and Therapeutics

Chapter 3
Application of Nanotechnology in Imaging and Diagnostics

Abstract The technological evolution and the knowledge gained to provide useful tools to the scientific community to develop molecular tools for toxic substance detection in everyday life and accurate determination of biomolecules (e.g., proteins) that can lead to diseases. Diagnosis at the molecular level was introduced to the clinical application to investigate diseases and to monitor their progress. The term nanotechnology on a chip refers to the application of molecular diagnosis in a variety of methods. The most accurate approaches that have been introduced to diagnostics and imaging are nanochips and nanoarrays. These approaches are sensitive and have greater speed that already exist in clinical applications. The term *theranostics* refers to the integration of the diagnosis/imaging and therapy approached. Nanotheranostics combine the simultaneous noninvasive diagnosis and treatment of diseases with the exciting possibility to monitor in real time drug release from the nanocarrier and distribution, thus predicting and validating the effectiveness of the therapy. Due to these features, nanotheranostics are extremely attractive for optimizing treatment outcomes in cancer and cardiovascular and other severe diseases.

Keywords Diagnostics • Imaging • Nanotheranostics • Biosensors • Microelectromechanical systems (MEMS) • Nanobiotechnology

3.1 Nanomolecular Diagnostics and Imaging

3.1.1 Introduction

During the past decades, there is great development in biotechnology and chemistry, as new chemical and biological molecules are both produced and used in various fields of human activities. The nonrational use of these products of high technology resulted in the promotion of problems from their use, such as the environmental pollution, the ecotoxicity, the need for the development of a biological cleansing unit for the chemical and biological waste, etc. Thus, the need for the development of new detection methods for these substances was essential. The bioanalytical assays that are based on protein or nucleic acid determination are valuable bioanalytical tools.

© Springer Science+Business Media Singapore 2016   65
C. Demetzos, *Pharmaceutical Nanotechnology*,
DOI 10.1007/978-981-10-0791-0_3

Table 3.1 Nanotechnological applications in diagnostics, imaging, and therapeutics

Diseases	Diagnostics, imaging, therapeutics
Cancer	Nanoparticle tracers and contrast agents for diagnosis
	Nanoparticles for monitoring of therapy
	Minimal invasive endoscope/catheter for diagnostics and therapy
	Nanostructured surfaces for biosensors
	New nanoformulations for targeting agents to tumors
	Implantable devices for localized delivery of drugs
	New therapeutic tools with physical mode of action
	Monitoring of therapeutic efficacy
Cardiovascular	Nanoparticles for theranostic approaches
	Implantable devices (nanosurface modification)
	Targeted drug delivery into plaques
Neuro degenerative	Image-guided implantation of advanced neurostimulators
	Semi-invasive nanodevices for drug delivery (for Parkinson)
	Nanoformulations for crossing the BBB
Metabolic (diabetes)	Encapsulation and monitoring of labeled islet transplants
	Whole body imaging of fat distribution with nanoparticles
	Implanted noninvasive continuous glucose monitoring
	Insulin measurement and delivery by nano-enabled devices
Inflammatory	Imaging of nanoparticle labeled with blood cells
	Soft nanomaterials for bone regeneration, rheumatoid arthritis, and Crohn's disease

Nowadays, the technological development and the extra knowledge gained helped scientists develop molecular tools for toxic substance detection in everyday life and accurate determination of biomolecules (e.g., proteins) that can lead to diseases. The clinical application of molecular technology diagnosis and the monitoring of human diseases are referred to as molecular diagnostics. The term relates to the expanded meaning of "DNA diagnostics" that is used during the nucleic acid, gene, and protein technology application. Monoclonal antibodies (MAbs) and enzyme-linked immunosorbent assay (ELISA) are also considered as diagnostic tools in *biotechnology diagnostics*. The term genomic diagnostics is used for applications in genomic molecular diagnostics that studies the organism genes, the base sequence, the structure, the mechanisms of regulation, the interactions, and the final medicines. Nanobiotechnology uses terms from molecular diagnostics that are included in the expanded category of biochips/microarrays, but the use of terms nanochips/nanoarrays is considered as a more accurate approach. Nanotechnology on a chip relates to various applications and performs assays in nano-volume capacities. A huge number of nanodevices that consist of organized nanostructures from inorganic materials are used to study the DNA sequence that has applications in medicine and biology. These nanodevices are characterized as markers.

We can mention nanodevices like sensors in nanodimensions that are accurate during assays, i.e., receptors and other living organism systems. Also nanoparticles that can be used as labeling systems can measure the presence or the efficacy of certain products with elevated sensitivity and greater speed in assays. Table 3.1 presents nanotechnological applications in molecular diagnostics, imaging, and therapeutics. It should be noted that biomaterials are considered as a breakthrough area of nanoscale technologies. Their properties provide new insights to the material science to produce innovative technological platforms for every day products, mainly in therapeutics and as daily goods [7].

3.1.2   Biosensors

The development in biosensor technology occurred due to the need of bioanalytical tool developments that have the ability to transform the interaction of the biological receptor with the molecule or the biomolecule that we need to define in analytical grade. The biochemical sensors are chemical sensors whose system is based in a biochemical mechanism. The fields in which biosensors technology is applied are many and continually developing during the years. Clinical diagnostics, biomedicine, quality control and pharmaceutical analysis, microbiology for virus and bacteria detection, food production and quality, veterinary, industrial waste control, industrial and toxic gas identification, and environmental pollution observation are some of the fields of biosensor applications. Their capability to grant assay results in low detection levels and high specialization in short time offered great advantages and benefits. The techniques used today for the analysis of biological and environmental samples usually take time, days or even weeks to offer a result. The delay on the start of a therapy or the application of the certain measures in environmental pollution puts a risk in human lives [18].

Also, in order to fulfill these assays, the use of special equipment and properly trained personnel raised the actual cost. Therefore, the application of sensitive and quick techniques has an advantage by limiting the result times and lowering the assay cost. Biosensors are the tools that have a biological substance (enzyme, antibody, nucleic acid, cells, or tissue parts) and in combination with a physicochemical analysis are used to detect molecules (Fig. 3.1).

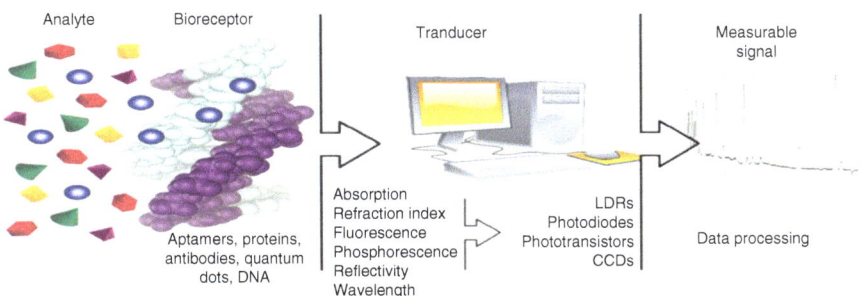

Fig. 3.1 Schematic representation of an optic biosensor (Adapted from [17])

Briefly, biosensors provide a reading after the reaction with the analytical sample without the addition of an extra reagent.

Each biosensor is composed from:

- A recognition biomolecule (sensitive biological element) that provides specialization to the sensor in relation to the sample (e.g., tissue, microorganism, organelle, receptor, enzyme, antibody, nucleic acid).
- A signal transducer (detector element) that acts in a physical/chemical way (optical, electrochemical, etc.). This part transforms the signal that is produced from the reaction of the sample and the biorecognition element, to a measurable signal.
- A signal transducer or an electronic processor for the result presentation.

The parts that the biosensor is consisted of and their combinations can contribute to greater sensitivity, specialization, and portability and improve the assay time needed. Usually the biosensor detection element that is responsible for the sample recognition is located onto the detection area surface and is immobilized. The advantages of this biomolecule immobilization are the following:

- The ability of continuous molecule detection in flow systems
- The use of well-defined biomolecule quantities
- The possibility of reusing the biosensor

The effective use of the biosensor is affected from the materials used, i.e., the materials used for the biosensor mechanical support, materials used as matrices, or membranes including the biological active factor. Biosensor support types include the inactive support for the mechanical support, the optical waveguide that plays an important role in sensor optical examination, and the active support like fluorescent nanoparticles, metal films, metals spheres, and inorganic and organic *micro-* and *nano*particles, and they function as small light sources and can take part in the spectroscopic plan. The size of the identified molecule is the one defining the biosensor's structural organization. For example, enzyme biosensors usually metabolize small- or medium-sized substrates. Hydrogel matrices are useful only for low molecular mass, molecules that can penetrate into the matrix and can be applied in biosensors with whole cells. There are three different test types where biosensors can be applied on:

- Direct type tests
- Indirect type tests with competitive link
- Indirect type tests with non competitive link

3.1.3   Nanocrystals/Quantum Nanodots

Quantum nanodots are spherical nanocrystals which have been used as diagnostic probes (imaging) and as therapeutic products [16, 19]. They can be produced from metal elements with semiconductor properties (e.g., CdS, CdSe, Cdte, ZnS, PbS) and other metals (e.g., Au). Quantum nanodots' diameter ranges between 2 and 10 nm that correspond to 10–50 atoms. Generally speaking, quantum nanodots consisted of a semiconductor core that is covered from a shell, e.g., ZnS to improve its optical properties and its solubility in buffers.

The use of quantum nanodots (or nanocrystals), semiconductor quantum dots (or semiconductor nanocrystals), and quantum dots (QD) that are composed from metal ions and colloidal stabilizers for biological labeling and imaging, used in medical applications, keeps on growing. The future applications include their use as fluorescent labels for cell labeling, intercellular biosensors for deep tissue and tumor imaging agents, and sensitizers for photodynamic therapy, while due to their surface chemistry, their use can be extended in biological applications because of their low cytotoxicity. Recently, there was a great interest due to their applications, for nanoparticle study used as nanoparticle-mediated drug delivery.

Nanocrystals (quantum nanodots) are designed with a few hundred atoms. Usually, they are consisted of II-VI or III-V group type atoms of the periodic system and have dimensions smaller than the Bohr's radius. Nanocrystals (Fig. 3.2) absorb light in a wide wavelength range, but they emit unicolor light in a wavelength that depends on the crystal size. Scientists can produce nanocrystals that will emit light in a specific wavelength, from infrared to ultraviolet.

Connecting nanocrystals with cells, proteins, and nucleic acids offers scientists the ability to observe one cell with a single nanocrystal, since the intensity of the light emitted is more than enough.

Nanocrystals are composed from inorganic structural elements; they are very stable, while their coating with dormant biocompatible material makes them less toxic compared with organic matrix. Targeting properties of QD can be achieved by modification of their surface properties. Polyethylene glycol coating can protect QD from MPS and reduce their toxicity after i.v. injection [16, 19]. Their binding with antibodies that correspond to some type of cancer creates exceptional advantages when used for diagnosis. After entering the organism, antibodies are attached to the nanocrystal, track the antigen target, and connect with it. Following, they receive radiation and emit light, allowing the detection of cancer tumor [1].

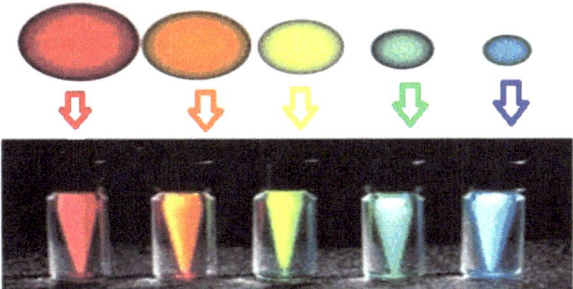

Fig. 3.2 Vials containing nanocrystals illuminated with light. Each nanocrystal emits light of a specific wavelength, even if illuminated with monochromatic radiation. The *colored spheres/ bullets* indicate the relative sizes of the nanocrystals

3.1.4 Nanoshells

Metal nanoshells [4], usually composed of gold, are a new group of nanoparticles with adjustable optical properties and can be used for biomedical applications. They are composed of a dielectric core covered by a metal shell. Silica and gold are the preferable biomaterials that are used for their production. They can convert their electrical energy into light in the spectra region from UV to IR which reflects to their optical changes that are considered important for their application as contrast agents [2, 7, 9]. The possibilities for simultaneous therapy and imaging of gold nanoshells, spherical nanoparticles with silicon dioxide core, and golden shells have drawn great deal of attention during the past two decades and have been highlighted as a very useful tool for cancer therapy and bio-imaging.

3.1.5 Carbon Nanomaterials

Carbon nanotubes (CNTs) with nanoparticles entrapped into graphite walls have been developed for biomedicine and pharmaceutical applications [12]. Carbon nanotubes present unique properties like great surface area, great mechanical strength, and extreme chemical and thermal stability. They can interact with cell membranes, biological molecules, and structural components inside the cell. A carbon nanotube is like a graphene lamellae enfolded in cylinder shape with a nanometer-scale diameter. It is a one-dimensional nanostructure and the ratio length to diameter is more than 10,000. In 1993 the first walled carbon nanotube was synthesized, while in 1991 the multi-walled carbon nanotube synthesis appeared. Sumio Iijima's report was of great value since it has drawn the scientific interest on this field [5, 6, 15]. Carbon nanotubes can be with multiple walls (multi-wall nanotubes, MWNT) with one central tube surrounded with one or more graphene layers or with a single wall (single-wall nanotube, SWNT) where there is only one tube and not additional graphene layers. When nanotubes are grouped together, they form nanotube bundles. In a nanotube bundle, commonly oriented unilamellar nanotubes are linked together with weak *van der Waals* forces, forming a two-dimensional triangular lattice. The coefficient of this lattice is 1.7 nm while the distance between nanotubes is 0.315 nm. NanoMix (Emeryville, CA, USA) is an in vitro diagnostic CNT sensor for monitoring functions of the respiratory system. This product is used to improve the sensitivity of the analysis and it is considered as a novel sensor [20].

3.1.6 Superparamagnetic Nanoparticles

Superparamagnetic nanoparticles have a metal core (iron, cobalt, or nickel) that is magnetically active and are used as contrast agents in magnetic resonance imaging. They have a diameter of less than 10 nm. Also, superparamagnetic nanoparticles have been coupled with antibodies, a fact that possibly allows simultaneous cancer imaging and therapy [7].

3.1.7   Inorganic Nanoparticles

The advantage of inorganic nanoparticles is the simple production process in mild conditions, with the desired size, shape, and porosity. Their very small size (less than 50 nm) gives them the ability to avoid the MPS, while pH changes do not affect them. Particles made from silica, argils, titanium, etc. are tissue biocompatible. Apart from that, monoclonal antibodies or other molecules can be connected on their surface for in vivo targeting. The growing interest to use inorganic nanoparticles, which recently characterized as self-therapeutics, in medicine is due to the unique size-, shape-, and morphology-dependent optoelectronic properties. Gold, silver, and platinum nanoparticles are developed for therapeutic applications with regard to cancer in terms of nanoparticles being used as a delivery vehicle as well as therapeutic agents. Inorganic nanoparticles show low toxicity and promise for controlled delivery properties, thus presenting a new alternative to viral carriers and cationic carriers. Inorganic nanoparticles generally possess versatile properties suitable for cellular delivery, including wide availability, rich functionality, good biocompatibility, potential capability of targeted delivery (e.g., selectively destroying cancer cells but sparing normal tissues), and controlled release of carried drugs. Critical properties of inorganic nanoparticles, surface functionalization (modification), uptake of biomolecules, the driving forces for delivery, and release of biomolecules are studied systematically.

3.1.8   Nanobars

Another nanodevice that allows cancer cell detection is composed of nanobars that carry antibodies and are attached from one side. When proteins enclosed in cancer cells connect with the antibodies attached to nanobars, nanobars start to slightly bend. This slight nanobar bending causes detectable potential change that is recorded as an analytical sign. Its evaluation leads to conclusions and information on the presence/absence of cancer cells, and the type of cancer cells, depending on the connection and the nanobar bending of the specific antibody.

3.1.9   Nanowires

Nanowires, 1–2 nm width, when placed in a suitable substrate can be covered with monoclonal antibodies against various cancer types. This results in sensitivity increase, approximately 100 times more than the existing diagnostic techniques, using the smallest sample quantity.

3.1.10   Magnetic Nanoparticles

Magnetic nanoparticles can connect to antibodies and enter the human organism. According to their ability of connecting with the antigen target that is overexpressed in specific cells, by applying an external magnetic field, nanoparticles lead to a certain magnetic behavior. In other words, the diagnosis is based on the external applied magnetic field change. Nanoparticles not connected with the antigen target are rotated freely and their total magnetic field is zero. On the contrary, nanoparticles connected with the antigen target cannot move freely and give a measurable magnetic field change. Many diseases can be detected simultaneously using magnetic nanoparticles with various antibodies. Magnetic nanoparticles improve the detection ability in lymph nodes of metastatic damages that are caused in breast or prostate cancer.

3.1.11   Microelectromechanical Systems (MEMS)

Microtechnology already affects the biological analysis field in a very important way. Evolutions on MEMS like functioning in a lab scale, on a completed chip, has created lab on chip, offering the possibility of substance detection (bioanalytical assay). The implementation of microcircuitry in devices helps in automation increase, creating advantages due to their mass production. BioMEMS have been designed to be applicable as sensors, e.g., to detect glucose levels in humans or in drug delivery [22]. The technique of changeable lithography used by microfluidics is being developed more easily especially in nanoscale than in microelectronics level. Scientists implementing nanofluids can work on smaller matter quantities and separate proteins or nucleic acids (DNA and RNA) according to size and shape. Nanomembranes present great evolution potentials on this field. For example, the single DNA or RNA beam's transit through a nanopore or a nanodiode forces the bases to pass through the specific nanopore, and then they are studied in polymorphism level (especially in conformational or structural polymorphism that is related to genetic diseases).

More in particular, conformational polymorphism is extremely difficult in studying, since the liquid crystal phase of the base plays an important role. It is difficult to determine and evaluate the base's *mesophase*, i.e., the change in the crystalline structure and not in the structural characteristics. These *mesophases* constitute conformational polymorphism that is above all a kinetic phenomenon. The change of the electric field where the nanopore is placed could be used in the identification of the passing by base. This way, its structural and the conformational integrity is detected, i.e., its normal state and smooth functionality. A possible outcome of this approach is the ability of placing consecutively the whole genome (the total genes of an organism) in just a few hours time.

3.2   Theranostics

The challenges of the next-generation therapeutics are the inhomogeneity and adaptability of major chronic diseases such as cancer and inflammatory disorders and the seemingly infinite variability of the patients' genomes. Facing the above challenges, the personalized (patient-tailored) medicine is the answer and adapts the treatment to the patient specific characteristics. This scientific orientation requires the knowledge of the molecular defects of the diseases and of the biological systems to effectively and efficiently identify those defects, deliver bioactive molecules, and monitor the therapy. The molecular imaging is a growing research discipline aimed at developing and testing novel tools, reagents, and methods to image-specific molecular pathways in vivo, particularly those that are key targets in disease processes. Personalized medicine is the ultimate goal of molecular imaging [13] which is an interdisciplinary approach for:

1. Identification of a marker of disease such as genes, cytoplasmic or free proteins, and enzymes.
2. A beacon (e.g., fluorescent or bioluminescent molecule, a metallic nanoparticle, an ultrasound air bubble) that can be attached to the marker via a chemical reaction or interactions.
3. Optical, ultrasound, electrical, or other methods can be used to identify the presence or absence of the marker and consequently of the beacon.

A diagnostic test that identifies patients most likely to be helped or harmed by a new medication (biomarker identification and validation, patient stratification, "predictive pharmacology") and targeted therapy and in vivo imaging based on the test (tests based on sophisticated technology involving genetics, molecular biology, and testing platforms such as microchips). The test results are used to tailor and monitor treatment, usually with a drug that targets and images a particular gene or protein.

Nanobiotechnology offers a promise to revolutionize the life sciences because it equips biologists with tools and materials that can interact directly with the biomolecules that they study on a daily basis biomolecules-material interaction, the *sine qua non* of nanobiotechnology. Both biotechnology and nanotechnology have matured to the point that their convergence offers opportunities for novel solutions to unmet needs in biology, pharmaceutics, and medicine. In the field of drug delivery, the particle is preloaded with a therapeutic agent, and, once at the target, the outer shell disintegrates, releasing the bioactive molecule. The medicine is released directly over the disease location at a high concentration, resulting in a more aggressive treatment fewer and much less intense side effects compared to current regimens, e.g., chemotherapy.

According to Kostarelos [8], the chemical character of the nanomaterial is evidently a fundamental parameter that contributes greatly to the biological responses obtained upon interaction with biological matter (i.e., modulation of the nanomaterial chemical character by covalent linkages can lead to dramatic changes in the physical and the biological properties of the nanomaterial; proof of concept that carbon nanotubes (see Chap. 3) can act as delivery systems for drugs, antigens, and genes with minimal cytotoxicity via a novel uptake mechanism which has proven to be energy independent) (see Chap. 2).

The term *theranostics* refers to the simultaneous integration of the diagnosis/imaging and therapy approached [10]. Therefore, the purpose is to diagnose and treat the diseases, when the diseases are most likely curable or at least treatable. Theranostic nanomedicine shows better characteristics than other theranostic agents since they have advanced capabilities, which include sustained/controlled release, targeted delivery, and imaging [3]. Nanotheranostics hold great promises because they combine the simultaneous noninvasive diagnosis and treatment of diseases with the exciting possibility to monitor in real time drug release and distribution, thus predicting and validating the effectiveness of the therapy. Due to these features, nanotheranostics are extremely attractive for optimizing treatment outcomes in cancer and cardiovascular and other severe diseases. The following step is the attempt to use nanotheranostics for performing a real personalized medicine which will tailor optimized treatment to each patient, taking into account the individual variability. Clinical application of nanotheranostics would enable earlier detection and treatment of diseases and earlier assessment of the response, thus allowing screening for patients which would potentially respond to therapy and have higher possibilities of a favorable outcome. This concept makes nanotheranostics extremely appealing to elaborate personalized therapeutic protocols for achieving the maximal benefit along with a high safety profile [14]. In an extensive review article by Zhang and coworkers, gold nanorods (AuNRs) are referred as promising for biomedical applications as they fuse in one system both imaging and therapeutic approach. AuNRs are well studied in in vitro experiments and in hyperthermia therapy, but efforts are needed in order to establish their effectiveness in therapeutics [21].

In other words, theranostic nanoparticles contain diagnostic and therapeutic functions in one integrated system, enabling diagnosis, therapy, and monitoring of therapeutic response at the same time. For diagnostic function, theranostic multifunctional nanosystem requires the inclusion of noninvasive imaging modalities. Among them, optical imaging has various advantages including effectiveness, sensitivity, real time, and convenient use and non-ionization safety, which make it the leading technique for theranostic nano-platforms [11]. For therapeutic function, theranostic nanoparticles have been applied to chemotherapy, photodynamic and photothermal therapy, and gene therapy.

3.3 Summary

Nanodevices are introduced to diagnostics and imaging as nanochips and nanoarrays are more accurate approaches providing sensitivity and greater speed in assays.

Theranostics refer to the combining of diagnosis/imaging and therapy approach in one nanodevice.

References

1. Akerman MA, Chan WCW, Laakkonenet et al (2002) Nanocrystals targeting *in vivo*. PNAS 99(20):12617–12621
2. Barton J, Halas NJ, Nest J et al (2004) Nanoshells as OCT contrast agent. Proc SPIE Int Soc Opt Eng 5316:99
3. Freitas RA Jr (2005) What is nanomedicine? Nanomedicine 1:2–9
4. Hirsch LR, Gobin AM, Lowery AR et al (2006) Metal nanoshells. Ann Biomed Eng 34(1):15–22
5. Iijima S (1991) Helical microtubules of graphitic carbon. Nature 354:56–58
6. Iijima S, Ichihashi T (1993) Single-shell carbon nanotubes of 1-nm diameter. Nature 363:603–605
7. Kewal KJ (2008) The handbook of nanomedicine. Humana Press, Basel, pp 36–37
8. Kostarelos K (2010) Carbon nanotubes. Fibrillar Pharmacol Nat Mater 9(10):793–795
9. Loo C, Lowezy A, Halas N et al (2005) Immunotargeted nanoshells for integrated cancer imaging and therapy. Nano Lett 5:709–711
10. Mura S, Couvreur P (2012) Nanotheranostics for personalized medicine. Adv Drug Deliv Rev 64:1394–1416
11. Murday JS, Siege RW, Stein J et al (2009) Translational nanomedicine: status assessment and opportunities. Nanomedicine 5:251–273
12. Pagona G, Tagmatarchis N (2006) Carbon nanotubes: materials for medicinal chemistry and biotechnological applications. Curr Med Chem 13(15):1789–1798
13. Rudin M, Weissleder M (2003) Molecular imaging in drug discovery and development. Nat Rev Drug Discov 2:123–131
14. Ryu JH, Koo H, Sun IC et al (2012) Tumor-targeting multi-functional nanoparticles for theragnosis: new paradigm for cancer therapy. Adv Drug Deliv Rev 64:1447–1458
15. Steinhart M, Wendorff JH, Wehrspohn RB (2003) Nanotubes à la carte: wetting of porous templates. ChemPhysChem 4:1171–1176
16. Stroch M, Zimmer JP, Duda DG et al (2005) Quantum dots spectrally distinguish multiple species within the tumor milieu in vivo. Nat Med 11:678–682
17. Shruthi GS, Amitha CV, Mathew BB (2014) Biosensors: a modern day achievement. J Instrum Technol 2(1):26–39
18. Thevenot DR, Toth K, Durst R et al (1999) Electrochemical biosensors: recommended definitions and classification. Pure Appl Chem 16(1–2):121–131
19. Voura EB, Jaiswal JK, Mattoussi H et al (2004) Tracking metastatic tumor cell extravasation with quantum dot nanocrystals and fluorescence emission-scanning microscopy. Nat Med 10:993–998
20. Wagner V, Dullaart A, Bock AK, Zweck A (2006) The emerging nanomedicine landscape. Nat Biotechnol 24:1211–1217
21. Zang Z, Wang J, Chen C (2013) Gold nanorods based platforms for light-mediated theranostics. Theranostics 3(3):223–238
22. Zuckermans ST, Kao WJ (2009) Nanomaterials and biocompatibility: BIOMES and Dendrimers. In: de Villiers MM, Aramwit P, Known GS (eds) Nanotechnology in drug delivery, vol 10. Springer/AAPS press, New York, pp 193–228

Chapter 4
Application of Nanotechnology in Drug Delivery and Targeting

Abstract Lipidic nanoparticulate self-assembled structures are effective carriers for drug delivery. This chapter describes the most famous nanotechnological drug delivery systems that are already used in clinical practice and clinical evaluation or in academic research. Liposomes are nanocolloidal lyotropic liquid crystals that are able to deliver bioactive molecules. Their membrane biophysics and thermodynamic properties reflect to the creation of metastable phases that affect their functionality and physicochemical behavior. Thermo- and pH-responsive liposomes are innovative nanotechnological platforms for drug delivery and targeting. Polymeric micelles and polymersomes are nanostructures that are promising drug carriers, while dendrimeric structures are considered as real nanoparticulate systems that are used in drug delivery and as nonviral vectors as well as in prevention of serious infections leading to diseases. Vaccines based on nanoparticles such as liposomes are an emerging technology and liposomes seem to meet the requirement criteria of adjuvanicity.

Keywords Packing parameter • Liposomes • polymersomes • micelles • Dendrimers • Vaccines

4.1 Pharmaceutical Nanotechnology

Nanotechnological systems are used as bioactive molecules' delivery systems for therapeutic purposes and for tissue imaging. Furthermore, new scientific studies combine nanosystem application for simultaneous disease treatment and monitoring disease progression, helping the clinical doctor to get the most accurate diagnosis. These systems are known as theranostics (diagnostic-therapeutic products) (see Chap. 3). In this chapter, the most important nanotechnological drug delivery systems will be studied, starting from their technological development in the laboratory to their clinical use regarding their biophysics, therapeutic efficacy, and safety. The pharmaceutical scientists have a great advantage when they try to understand the behavior of these nanosystems in a technological point of view and by correlating their functionality with the living organisms as well. Pharmaceutical nanotechnology research aims in the development of new medicines based on bioactive molecule

© Springer Science+Business Media Singapore 2016
C. Demetzos, *Pharmaceutical Nanotechnology*,
DOI 10.1007/978-981-10-0791-0_4

delivery with nanosystems and in geometric increase of medicines that are currently in clinical studies. Pharmaceutical industry follows the same path, investing in the bioactive molecule development, in pharmaceutical nanotechnology point of view.

4.1.1 Lipidic Nanocarriers

Lipidic nanocarriers [4] are considered the most beneficial and promising technological platforms for encapsulating bioactive molecules. The biocompatibility and biodegradability due to their lipidic nature are advantageous properties, among others, like high ability entrapment of bioactive molecules with different solubilities. However, they have emerged as interesting nanosystems for delivering bioactive molecules to the target tissues. Their ability to improve the active and passive targeting mechanisms is of importance for exploring new disease targets. The most common and interesting lipidic carriers are liposomes. Commercially available lipidic nanocarriers that contain amphotericin B in the market are Abelcet® (Sigma-Tau PharmaSource Inc.) which is a lipidic complex suspension for intravenous infusion. It consists of dimyristoyl phosphatidylcholine (DMPC) and dimyristoyl phosphatidylglycerol (DMPG) in a 1:1 drug to lipid ratio. It forms ribbonlike particles with a size of 1.6–11 μm. Amphotec® (Ben Venue Laboratories, Inc, Bedford, OH) consists of amphotericin B in a complex with cholesteryl sulfate at a 1:1 drug to lipid molar ratio. The shape of the particles is determined as disk with a size of 120–140 nm. Two more lipidic nanocarriers are also referred in the literature. These lipidic nanosystems are the archaeosomes and nanocochleates.

Archaeosomes are lipidic bilayer vesicles composed of ether lipids that are obtained from the bacteria *Archaea* (Archaebacteria). They are more stable than liposomes to the changes of temperature, pH, oxidation, and hydrolysis. Omri and coworkers have reported the i.v. and oral delivery of archaeosomes in mice [62].

The name cochleates first presented in the pioneer work by Papahadjopoulos and coworkers [64]. The addition of divalent cation such as Ca-2+ (10 mM) to the lipidic preparation composed of phosphatidylserine in aqueous NaCl buffer following incubation for 1 h at 37 °C leads to the production of multilamellar structures which present cylindrical morphology. The spiral configuration was studied by using freeze fracture electron microscopy. Papahadjopoulos and coworkers proposed the name cochleate lipid cylinders. Nanocochleates are lipidic nanocarriers composed mainly of charged lipids and a divalent cation. They are applied for delivery of DNA, proteins, and peptides. Their structure (cigar-shaped MLV structure) is the reason for considering them as candidates for oral and systemic delivery of bioactive molecules [4].

4.1.1.1 Liposomes

Just before mentioning lipid nanosystems for therapeutic and diagnostic purposes, like liposomes, a very important step is the short description of the lipid bilayers that consisted of. Liposomal lipid bilayer is a study subject of many scientists like physicists, chemists, biologists, biophysics, etc., while its thermo-dynamic behavior is also an important study subject that relates with their stability and functionality. Its physicochemical characteristics and thermodynamic phase transition (see Chap. 2) that characterize the lipid system thermotropic behavior and, therefore, its effectiveness are related to the geometry of its structural materials. The question arising during lipid membrane study is: "which is the driving force that causes the self-assembly of biomaterials, e.g., lipids – and in the case of liposomes, mostly phospholipids – in structures, e.g., spherical structure instead of flat (foliate) structure?" or even better "which is the driving force creating the specific nanostructure type?" The answer is based on the physicochemical and thermodynamic characteristics of the medium where a nanostructure is self-assembled.

The geometric characteristic determination of nanoparticulate systems containing amphiphilic surfactants (e.g., phospholipids) such as liposomes is based on the packing parameter S_p. This parameter is symbolized in this book as S_p so there will be no confusion with parameter S that will be mentioned later on and relates to the biomolecule orientation in a nanosystem (Figs. 4.1 and 4.2) [39]:

$$S_p = \frac{V}{a_0 l_c}$$

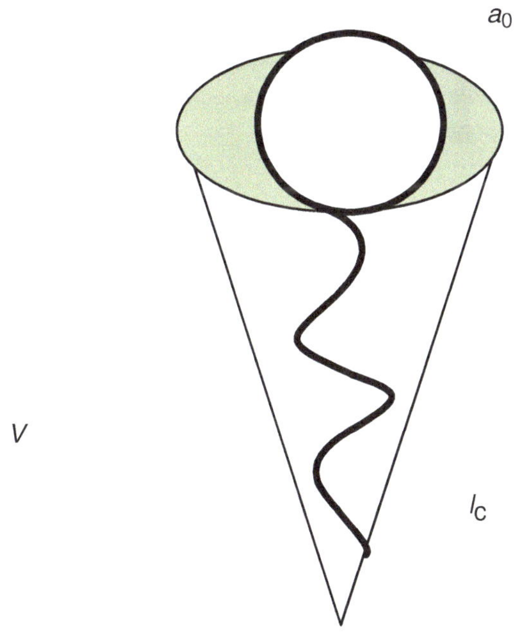

Fig. 4.1 The geometric characteristics of surfactant amphiphilic

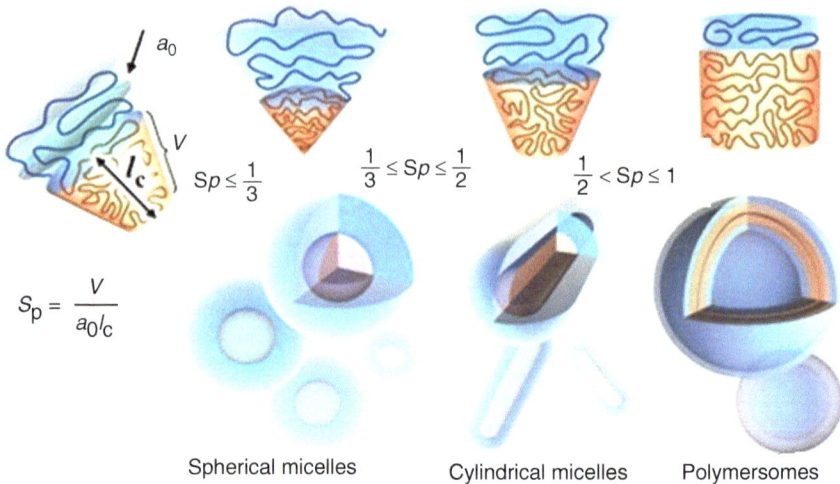

Fig. 4.2 Nanostructures of self-assembly of biomolecules resulting in a different shape. Calculation of packing parameter, S_p

$$S_p = \frac{V}{a_0 l_c} \tag{4.1}$$

where V is the volume of the hydrophobic chain(s), l_c is the length of the hydrophobic chain(s), and a_0 is the surface area of the hydrophilic polar group/head of the amphiphilic (surfactant) molecule, e.g., phospholipids. The geometrical characteristics of the structural units of nanosystem affect through their self-assembly process, its organization, shape, and structural characteristics. Another important parameter is the one referring to the biomolecule orientation regarding the vertical settlement axis. This parameter is characterized with the letter S and can have values 1 and 0.3–0.9 that have been emerged from the experimental data.

The mathematical type that calculates the S and based on its value classifies the system in crystalline form ($S=1$) in isotropic liquid ($S=0$) and in liquid crystal ($S=0.3$–0.9) is the following:

$$S = \frac{3\cos^2 \theta - 1}{2} \tag{4.2}$$

$S=1$ (crystal), $S=0$ (liquid), and $S=0.3$–0.9 (liquid crystal). Where $\cos^2$ is the settlement axis angle of the nanosystems' biomolecule to the vertical axis. Important categories of self-assembled nanosystems with variations regarding the biomolecule chemical structure will be presented in detail in the following pages.

Liposomes were firstly described by Alec Bangham in 1965 [5]. In reality, the simple phospholipid stirring (Fig. 4.3) in water has led to the development of a phospholipid dispersion that aimed in cell membrane study regarding its biophysical

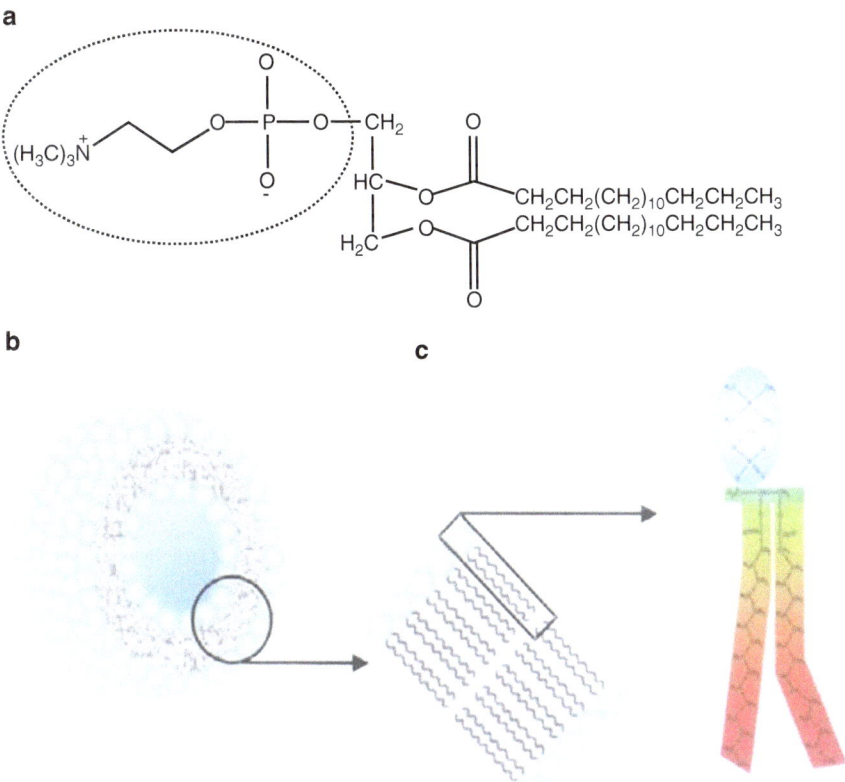

Fig. 4.3 (**a**) Structure of phospholipid. In cycle, the polar group (choline). (**b**) Liposomes (**c**) lipidic bilayer phospholipid

behavior. Liposomes belong to the self-assembled colloidal nanodispersion systems composed of lipidic bilayers that are able to incorporate lipophilic bioactive molecules or to entrap them into their aqueous core [5, 65]. This phospholipid dispersion was basically related to large phospholipid structures of irregular shape and micrometer size with multiple lipid layers. The lipid layers' polar heads were oriented toward the water molecules where the lipid chains have created the lipid bilayer through hydrophobic interactions. This effort was the beginning of the nanosystem development, liposomes that delivered bioactive molecules. During the past decades, liposomes are one of the most promising bioactive molecule delivery nanosystems. Liposomal technology is a developing research field, while scientific reports in this field are hundreds per year. Liposomes are closed *pseudo-spherical* structures that consisted of one or more lipid bilayers where inside them entrap water media and are characterized as thermodynamically unstable colloidal dispersions. Lipid bilayers consisted mainly of phospholipids and cholesterol without excluding the use of other biomaterials, e.g., polymers as liposomal structural units. Phospholipids, due to their amphiphilic character, in water environment orient so their polar heads are toward the water medium, while lipophilic hydrocarbon chains are protected from the water molecule connection, developing hydrophobic interactions. On liposomal surface, small molecules or macromolecules can be connected; these include antibodies that change the physicochemical properties of their surface. This fact is very important and depends upon the nanosystem functionality and its physical stability on time of production and over time.

The discovery of multilayer's irregular structures that lipids form where in water medium could be of use as a simulation model for cell membrane is a historical observation that enriched the biophysical and colloidal dispersion system science. In the early 1960s, there was a great interest in studies related to lipid colloidal system properties that could be used as design standards of cell membrane behavior. Initially they were called bangosomes, since AD Bangham was the first one that noticed them, but later on they were called liposomes from the Greek words λίπος (fat) and σώμα (body) [5]. Bangham had suggested the term amphisomes as more appropriate since the cell membrane structural parts were amphiphilic molecules. Studying the cell membranes' dynamic properties like fluidity and lipid layers' mesophase change was a challenging field when studying biophysics and thermotropic behavior of lipid membrane and, therefore, liposomal dispersions. Lipid vesicles in their initial form (simple dispersion in water environment) had no thermodynamically organized structures and were characterized as multilamellar vesicles (MLVs).

In 1967 Demetrios Papahadjopoulos and coworkers [63] described the structure of sonicated microvesicles that later on were known as small unilamellar vesicles (SUVs). These researches offered the biomembrane potential structure, explaining lipid layer fluidity and diffusion properties. The evolutions in this scientific field have created a fertile ground in scientific groups that started studying the possibility of applying lipid dispersions as vehicles for bioactive molecule transfer and delivery. During the past years, research related to liposomes as transport and delivery vehicles for therapeutic purposes has been developed [41]. The scientific interest in this area included the interaction of encapsulated biomolecules (nucleic acids, proteins, bioactive molecules) with the liposomes' structure units, their in vivo administration, the mechanisms when liposomes enter the cell, and their immunological properties. Changing the lipid composition (phospholipids and generally lipids of different structure), ζ-potential, liposomes' size, and size distribution were characterized as important properties that could define liposome behavior in vivo and in vitro. Phospholipids' distribution in water medium resulted in spontaneous development of lipid vesicles, and not liposomes, since liposome development needed a specific process (see Appendix). Lipidic vesicle structures (e.g., double layer, cubic structures, hexagonal structures, etc.) depend upon the phospholipid or lipid concentration that takes part in the process and their geometry that is related to their chemical structure. If the conditions mentioned before, i.e., phospholipid/lipid concentration and chemical structure, lead to lipid bilayer development, then the hydrophobic parts come in contact with water environment in a non-favorable energy way. This results to a bending so they will interact through hydrophobic interactions minimizing lipid chain exposition in water medium. The energy required for lipid bilayer bending is covered from the energy increase due to lipid bilayer's hydrophobic section exposition to the water environment. The result of hydrophobic lipid chains' bending and assemble is the development of multilayer lipid vesicles. Their description was reported for the first time in 1965 from Bangham and his coworkers [5]. So, lipid vesicles are the closed *pseudo-spherical* structures of one or more lipid bilayers that entrap the dispersion medium where they are. This formation offers lipid vesicles the property of being consisted of both the hydrophilic dispersion medium and the lipid chain hydrophobic section, respectively. For this reason, they can entrap hydrophobic, amphiphilic, or even hydrophilic bioactive molecules. Liposomal production with a specific process, mentioned in the Appendix – i.e., the lipid vesicle production with structural organization depended on the biomaterials' structural and geometrical characteristics, physicochemical characteristics, and their energy content – is an important fact directly related to their physical stability, the bioactive molecule entrapment ability, their release rate, and their efficacy according to their ADME (absorption, distribution, metabolism, and excretion) profile that will develop the final pharmaceutical product. During the past decades, liposomes are one of the most promising bioactive molecule carriers and delivery systems to the damaged tissues, and liposomal technology is a potential research development field while the scientific publications in this field are hundreds per year. According to the above, it can be said that liposomes are closed *pseudo-spherical* thermodynamically unstable lipid structures of dispersed lyotropic liquid crystals that are composed of one or more lipid bilayers that can bind on the inside of the water medium where they are. Lipid bilayers are composed of phospholipids, cholesterol, or other lipid molecules. Cholesterol is an important biomolecule that was chosen among the others through centuries in natural evolution process, since the system demanded thermodynamic *self-sufficiency*, structural stability, survival time ability, and life perpetuation through multivariate and multilevel cell organization. Cholesterol has been chosen as the most appropriate biomolecule that can sufficiently regulate the lipid bilayer fluidity that is necessary for its functions [38, 53]. Also, on liposomal surface, small molecules or macromolecules can be attached and modify their surface properties, for example, targeting antibodies through antigen connection. Nowadays, liposomes are considered as nanoparticles that develop liposomal colloidal nanosystems, extremely useful and highly promising for new pharmaceutical product development [28].

The target of each pharmaceutical formulation is the distribution optimization of the bioactive molecule in the human organism, the control over its release rate, and the maximization of the therapeutic result. Liposomes have the ability to entrap and transfer a wide range of bioactive molecules and have the following advantages:

- Liposomes can direct the bioactive molecule in specific parts of the organism (damaged tissues) enhancing the therapeutic effect while minimizing its accumulation in healthy tissues and, therefore, limiting toxicity and side effects. Targeted therapeutics is a continuously developing research activity.
- Liposomes can act as transport and delivery vehicles and as bioactive molecule accumulation tanks. Liposomes release the bioactive molecule in a controlled rate, modifying the bioactive molecule's pharmacokinetics.

- Protection of the entrapped bioactive molecule physicochemical integrity. Bioactive molecules entrapped in liposomes are protected since there are no enzymatic degradation processes taking place.
- Liposomes have the ability of transferring bioactive molecules inside the cell through different mechanisms, like fusion, phagocytosis, etc.

Moreover, the physicochemical properties and their behavior in biological media promote advantages over other delivery nanosystems. We have to refer that thermodynamics which relates to the stability profile of liposomes should be kept into consideration. It is worthy to note (see Chap. 2) that liposomes are not at thermodynamic equilibrium, but they behave as a kinetically trapped nanosystem, contrary to those nanosystems (e.g., polymeric nanosystems or microemulsions) that they are affected when changes in their environment occurred. However, liposomes preserve their physicochemical properties that promote their usefulness as drug delivery nanosystems [42].

Nowadays liposomal products are in the market for therapeutic purposes. Due to their double nature (lipidic bilayer, hydrophilic inside), liposomes can be used as carriers for both lipophilic and hydrophilic bioactive molecules. Depending on their nature, various bioactive molecules can be placed into the lipid bilayer or in the hydrophilic section of the liposome. In the first case, the bioactive molecule is incorporated into the bilayers of liposome, while in the second case, the bioactive molecule is encapsulated into the aqueous interior of liposome. Bioactive molecules entrapped in liposomes seem to be more effective to treat diseases and to protect healthy tissues from their exposure in toxic biomolecules. Liposomes are able to transfer infected organism antigens, malaria antigens, and bacterial toxins, and these have been successfully used for the development of chemical or cellular immunity in test animals. All of the above show that liposomes can be protein and peptide transferring systems but more research is needed for vaccine development.

Liposomes could be considered as artificial biomembranes, and their relationship regarding the biophysical behavior between biomembranes and cell biology has promoted them as leading nanosystems to deliver drugs to the target tissues. Liposomes have been successfully used in the field of cell physiology, and the pioneer work of Gregoriadis and coworkers in the 1970s provided evidences regarding the use of liposomes in enzyme replacement therapy [25–27]. They can be developed having a specific composition simulating cell membrane functionality, offering a model system for their study. Due to their relevant structure with the basic cell membrane components, they have been used to both stimulate and study them. These studies are on their thermotropic behavior, i.e., on the phase transitions, due to heat absorption or elimination. This thermotropic behavior can be related with the cell behavior and many assumptions can be concluded on their functionality. Also, this relation can highlight the *disease factors* and the biophysical changes due to *mesophases* resulting from phase transitions. The term *disease factor* can be related to the cell biophysical and thermodynamic behavior, e.g., a microbial infection can cause a cell membrane biophysical functionality change and, therefore, act as a *disease factor*. It is obvious that the more cell membranes and other cell organelles, even genetic material's phase transitions, affect the living functions, the more dangerous the *biophysical disease factor* becomes for the organs' function and for life itself. Liposomes have been used successfully for bioactive molecule action mechanism investigation, i.e., aminoglycoside antibiotics' cytotoxicity mechanism. Also, they have been used to understand the local anesthetic mode of action and as models for studying the phototoxic active oxygen-mediated incidences. Table 4.1 presents liposomal applications in various scientific fields. It should be mentioned that liposomal applications in medicine and pharmaceutics are due to the need of new medicine development, especially new bioactive molecule delivery moieties with specific physicochemical characteristics. Liposomal properties related to their biocompatibility with cell membranes, their ability to transfer and deliver bioactive molecules and biological products, and their ability to target damaged tissues connecting with macromolecules on their surface (e.g., antibodies) are the basic reasons for health science applications.

Most important parameters studied in liposomal technology are:

- Liposome lipid composition. The predominant lipids are phospholipids and cholesterol.
- Temperature, pressure, ionic strength, and presence of ions or macromolecules, e.g., proteins, enzymes, and bioactive molecules in water environment where liposomes are.
- Lipid membrane permeability, elasticity, width and shape, and its ability to interact with other cell membranes.
- Lipid concentration in the final dispersion system.

The classification of various liposomal types can be done according to size and number of bilayers. According to these criteria, they can be distinguished in:

- Multilamellar vesicles. They consisted of many concentric bilayers of 500–5000 nm size. These can be also divided in large multilamellar vesicles (MLVs) that have up to 20 bilayers and in oligolamellar vesicles (OLVs) that usually consisted of 5 lipid bilayers.
- Large unilamellar vesicles (LUVs). They have only one bilayer and their size is between 200 and 800 nm.

Table 4.1 Applications of liposomal technology

Science	Application
Mathematics	Topological study of space of two- or three-dimensional surfaces. Fractal geometry
Physics	Physical properties of biomaterials
Biophysics	Permeability, transitions of liquid crystalline phases of materials
Physical chemistry	Study of physicochemical characteristics of colloidal systems
Biology and biochemistry	Biological membrane models. Studies on the interactions between cells
Pharmaceutics	Drug delivery and targeting
Medicine	Diagnostics, pharmacological effectiveness of liposomal products

- Small unilamellar vesicles (SUVs). They have only one bilayer and their size is 100 nm.
- Multivesicular vesicles (MVVs). They include many sections in their interior, but contrary to MLVs, they are not concentric.

An additional liposomal category can be according to the lipid bilayer composition, and therefore, there are the following:

- Conventional iposomes. Conventional are called the liposomes that consisted of neutral phospholipids (neutral phospholipids of ionic nature – their total charge is zero as they have equal number of positive and negative charges).
- pH-sensitive liposomes. These liposomes mainly consisted of phospholipids, e.g., DPPE 1,2-dipalmitoyl-sn-glycero-3-phosphatidylethanolamine. According to Connor and coworkers [16], pH-sensitive liposomes composed of DPPE promote fusion, while the presence of phosphatidylcholine inhibits the fusion process according to water environment pH that it is exposed to, they present a different charge [46].
- Cationic liposomes. Cationic liposomes are those with a positive charge [e.g., DOTAP 1,2 dioleoyl-3-(trimethylammonium)-propane, DOTMA chloride N-{1-(2,3-dioleoyloxy)-propyl}-N,N,N-trimethylammonium]. They are used exclusively in intracellular transfer of negatively charged macromolecules like nucleic acids (DNA and RNA) and other oligonucleotides.
- Immunoliposomes or antibody-targeted liposomes [33]. This category relates in liposomes of targeted delivery that have on their surface an antibody that functions as a detection center for surface antigens of target cells. Usually, the antibody is covalently bonded with the active group on the liposome bilayer surface, usually a maleimide molecule.
- Long-circulating or sterically stabilized liposomes. These liposomes include phospholipids in their bilayer. These phospholipids are covalently bonded with polyethylene glycol (PEG). PEG offers a kind of steric stabilization, a barrier for interactions between liposomal surfaces and biological environment's components. These interactions usually include plasma protein penetration into the bilayer that can destabilize the liposome structure or have opsonization properties, whose interference lowers the immune system phagocytes' recognition ability. So, these liposomes stay for a long time in plasma circulation without having a small size.
- Stealth liposomes. These liposomes' structure is modified in order to lower their destruction rate from macrophage cells and to maximize their stay in the organism. Usually, they have a small size and on their surface they have polymers like polyethylene glycols (PEG), carbopol, and polyvinyl alcohol. Liposomes can act as bioactive molecule "tanks" for their extended release.
- Liposomes with macromolecules on their surface. These liposomes, known as ligand-targeted liposomes (LTLs), can have on their surface monoclonal antibodies, peptides, polysaccharides, receptors, hormones, vitamins, growth factors, etc. Macromolecule connection on liposomal surface can be either directly or through

a "ligand." In the case of ligand-targeted liposomes, bioactive molecules are connected on polymer chains (e.g., PEG).

Depending on macromolecule type on the liposome surface, these can be classified into the following:

• Liposomes with peptides on their surface. In this category, there are liposomes that have plasminogen on their surface and they target fibrin clots. Also, there are peptide liposomes that block angiogenesis in endothelial cell targets.
• Liposomes with polysaccharides on their surface. These liposomes through their surface polysaccharides are directed to specific cell targets.

Moreover, liposomes can be used in imaging when loaded with the appropriate contrast agent. Several efforts have been made for the accumulation of contrast agents in the required area by using carriers of the agent. Liposomes were found to be the most relevant nanoparticulate nanosystems because of their release of the agent profile [85]. Their applications in imaging and as diagnostic agents are mainly based on their biocompatibility and on their entrapment efficiency of diagnostic agents [29, 39]. Liposomes are attractive carriers that are able to overcome skin barrier and depending on their composition can behave as penetration enhancers mainly that with high values of hydrophilic-lipophilic balance. Transfersomes are liposomal formulations that have been introduced by Cevc [12] and can promote great fluxes of active ingredients through the skin. More innovative products are produced for cosmetics based on liposomal concept delivering compounds such as antioxidant, hydrated agents, etc. and even proteins [88]. It should be mentioned that, not surprisingly, liposomal technology accommodated nonmedicinal and pharmaceutical areas such as catalysis, cosmetics, ecology, etc. [40].

Initially, the bioactive molecule is dispersed in the multilayer lipid vesicle environment that is developed during lipid dispersion in aqueous medium [51, 52]. This dispersion depends on the physicochemical characteristics of the molecule and the vesicles formed. The presence or absence of charged groups within the bioactive molecule that are created depending on the pH of the dispersion medium is a crucial parameter in bioactive molecule encapsulation process in liposomes. Liposomes will be developed according to the multilayer lipid vesicles and their membrane physicochemical characteristics. Regardless on the liposome production method that follows the initial lipid dispersion in water medium and the development of multilayer vesicle structures, the bioactive molecule encapsulation into the liposomes that will be developed later on is based on three different techniques. The technique to be used is chosen according to the bioactive molecule physicochemical properties that can be either hydrophilic, hydrophobic, amphiphilic, or amphoteric (weak base or acid).

• Encapsulation. Encapsulation as a term is used basically in case of hydrophilic bioactive molecules. The process involves the bioactive molecule phospholipid or other lipid hydration with water medium. During the multilayer lipid vesicle development, the hydrophilic molecule is passively encapsulated into the inner water environment and between the bilayers. Following, liposomes with the bioactive molecule encapsulated on the water environment are developed according to process described in the Appendix of this chapter.

- Incorporation. A hydrophobic or amphiphilic molecule is dissolved in organic solution and then mixed with phospholipids or lipids. Evaporation follows on rotation device in vacuum and a thin lipid film is produced. The thin lipid film, phospholipid and/or lipid, bioactive molecule mixture hydration with the water solution results in multilayer lipid vesicle formation where hydrophobic or amphiphilic bioactive molecules are encapsulated in their lipid bilayer due to hydrophobic interactions with its lipids. Liposomes are then developed according to methods described in the Appendix of this chapter.
- Active loading. It is based on bioactive molecule passive diffusion through liposome lipid bilayer. Molecules like weak acids are in neutral or ionic form depending on the pH of the dissolution media. A bioactive molecule like that can penetrate the liposomal bilayer. The pH inside the liposomes is regulated so the bioactive molecule can be in its ionic form and due to charged groups cannot penetrate the lipid bilayer barrier and, therefore, accumulates inside the liposomes. This method includes initially, the liposome development through multilayer lipid vesicle development and then bioactive molecule addition through encapsulation in liposome water environment as mentioned earlier. An example of bioactive molecule active loading in liposomes is the encapsulation of the anticancer bioactive molecule, doxorubicin.

According to Sidone and coworkers [73], the pharmacokinetic variability of liposomal agents was 2.7-fold and 16.7-fold greater than non-liposomal agents as measured by ratio of AUC CV% and ratio of AUC_{max} to AUC_{min}, respectively. It is of importance to figure out that the incorporated bioactive molecule into liposomes does not interact with the site tissue until it releases from the liposomal formulation. It is obvious that the pharmacokinetic profile of the liposomal product combines the pharmacokinetics of the liposomal delivery system as well as the pharmacokinetics of the incorporated bioactive molecule after it releases from the carrier. However, the total pharmacokinetic profile of a liposomal product still remains a scientific area to be developed, because the rate of the release affects the overall pharmacokinetics of the liposomal product [1]. Today many anticancer liposomal formulations are available in clinical practice, while a huge number are in clinical trials. The first one which received approval in 1995 was doxorubicin HCl as liposomal injection (Caelyx[R] in Europe and Doxil[R] in the United States), to treat HIV-associated Kaposi sarcoma and ovarian and breast cancer [75]. The published data by Allen and Stuart [1] showed that the pharmacokinetic data between the two well-known anthracyclines, i.e., doxorubicin and daunorubicin incorporated into different composition liposomal formulations with those of the free form of the two anthracyclines, showed that the clearance rates of the liposomal formulations are lower than that of the free form of the drugs. It is of importance that the sterically stabilized (surface PEG grafted) liposomal formulation that corresponds to Doxil/Caelyx has shown a decrease clearance rate, in comparison with that of DaunoXome® which is the liposomal formulation of daunorubicin, even though the liposomal vehicles of DaunoXome have larger hydrodynamic diameter (size).

Thermo- and pH-Responsive Liposomes

The development of stimuli, dual- and multi-stimuli-responsive nanosystems, follows the same approach for improving the pharmacotherapy as stealth and *chimeric* liposomes (see Chap. 5) (combining two different or same biomaterials). These systems use intrinsic or extrinsic (external) stimuli as triggers in order to succeed in site-specific drug delivery and to improve the safety and efficacy of bioactive molecules. pH-, thermo-, redox-, enzyme-, magnetic field-, and light-responsive nanopreparations have been already studied extensively [13, 20]. Temperature-sensitive drug delivery systems offer great potential over their counterparts due to their versatility in design, tunability of phase transition temperatures, passive targeting ability, and in situ phase transitions [37]. Thermosensitive liposomes are developed in order to improve tumor accumulation, trigger liposomal drug bioavailability, enhance drug delivery specificity (i.e., *drug painting dosing* and *chemodosing*) and drug internalization, and personalize the treatment [77, 91]. Thermosensitive liposomal formulations minimize the toxicity of the encapsulated active substance/anticancer agent, control the release rate, and enhance the long-circulating properties by modulating the composition and the "smartness" "decision making" of the nanocarriers [44]. ThermoDox® (Celsion) is a nanoengineered drug delivery system in clinical trials (phase III) and is a low-temperature-sensitive liposomal formulation incorporating doxorubicin (anthracycline) for the treatment of metastatic malignant melanoma and liver cancer. ThermoDox is the first heat-activated liposomal formulation, which consisted of three synthetic low phase transition temperature phospholipids and releases the anticancer agent at 39.5 °C [22]. In the past decades, there have been reported and developed many strategies for formulating pH-sensitive liposomes. The main categories based on the components and the mechanisms of triggering pH sensitivity are listed below. The first way is to combine polymorphic lipids such as phosphatidylethanolamine (PE), diacetylenic phosphatidylethanolamine (DAPE), dioleoylphosphatidylethanolamine (DOPE), and palmitoyl-oleoyl-phosphatidylethanolamine (POPE) [74]. Secondly, the majority of these systems may contain "cage" lipid derivatives, such as N-citraconyl-dioleoylphosphatidylethanolamine (C-DOPE) and N-Citraconyl-dioleoyl-phosphatidylserine (C-DOPS) [46]. The mechanism of action of pH-sensitive liposomes is presented in the literature [47, 74]. Since pH-sensitive liposomes cannot always sustain a slow and steady release, especially in a physiological pH solution, thus resulting in cytotoxicity for normal tissue, more sophisticated structures are being investigated, in order to overcome this possible instability of pH-sensitive liposomes. The investigation in the area of pH-sensitive liposomal vehicles for drug delivery is still an interest approach, and because of the new intracellular targets that are recognized, the efforts in this scientific field should be expanded [15].

Immunoliposomes

Immunoliposomes belong to the lipidic class of carriers that are able to reach to the target tissue via the active targeting process. They possess macromolecules on their surface such as antibodies, carbohydrates, and hormones that act as detection centers from surface antigens that are on cell targets. During the 1980s, various techniques connecting monoclonal antibodies on liposomal surface have been developed. The preferable strategies by which the macromolecules attached on the outer surface of the liposomal membrane are adsorption to the outer surface of liposomal vehicles, insertion into the lipid bilayers, via biotin-avidin pair, and finally covalent binding [39]. The macromolecules that they are attached on the surface of liposomes have complementary ligands on the target cells, and consequently when they arrive close to the target cell, liposome binds specifically to the target site. Bioactive molecules encapsulated in liposomes that have monoclonal antibodies on their surface include anticancer agents like doxorubicin (anthracyclines), vinca alkaloids, and taxol (taxanes) that are released in the surrounding area of the target cell, reducing the adverse drug reactions and the toxicity and improving the therapeutic effect. The research on this medicines aims to the pharmacokinetic parameter improvement and the increased concentration of the bioactive molecule on target cells using immunoliposomes [33].

Mitochondriotropic Liposomes

Particulate delivery systems of bioactive molecules in highly specific targets such as subcellular organelles represent a great challenge in drug targeting. The mitochondrion is an organelle composed of two membranes which create two separate compartments and is responsible for the energy metabolism. They are unique organelles as they contain their own genome mitochondrial DNA (mtDNA). It is well known that possible drug targets are located inside the mitochondrial matrix and it is of particular interest to design and develop carriers that are able to move through the mitochondrial matrix to release drug to the target. *Mitochondrial Medicine* is a new field of biomedical research. Mitochondriotropic liposomal approach is considered as an effort for producing mitochondria-targeted particulate drug and DNA delivery systems. The use of reconstructed proteoliposomes containing mitochondrial membrane components was an effort to prove the hypothesis that liposomes by modifying their surface with a mitochondriotropic residue could be rendered as mitochondriotropic [19].

Analyzing Liposomes

Liposomes can be used as carriers for bioactive molecule (hydrophilic, hydrophobic, amphiphilic macromolecules and genetic material) encapsulation/ incorporation or transport. This property seems to be due to two different areas: a polar interior cavity and a lipophilic bilayer. Hydrophilic groups in the outside liposomal surface develop macromolecular attachment places like antibodies, peptides, etc. These properties have led scientists in the use of liposomes in bioactive molecule-targeted delivery research and in the field of diagnostics. The basic liposomal application as analytical tools is related to the immobilization of the label molecule on the liposome surface and to the entrapment of the marker molecule inside the cavity. The marker allows liposome detection and quantitative determination and, therefore, detection and quantitative determination of the molecule under assay, using the appropriate analytical method (optical or electrochemical).

Liposomes as analytical diagnostic tools have the following advantages:

- Great outside surface, where various biological macromolecules can be attached, e.g., antibodies and peptides.
- Flexibility when choosing the individual components. This fact allows the production of liposomes with desired properties, required stability, and connection ability with various targeted/detection molecules.
- Great internal volume that allows the entrapment of a huge number of hydrophilic indicator molecules like dyes.
- Time-independent signal production.

The use of liposome flow injection systems is mentioned since 1988. Flow injection systems offer advantages like greater accuracy and possibility of automotive process. This method has been used in theophylline, estrogen, fumonisin B1, *E. coli*, etc. detection.

Biosensors According to Liposome Technology

Bioanalytical assays related to quality and quantity detection of various substances into biological fluids (e.g., blood) can be classified into two broad groups: those depending on protein detection and those depending on nucleic acid detection. Examples of assays depending on protein detection are enzyme-linked immunoassay (ELISA), radioimmunoassay (RIA), and immunoblotting. These assays depend on antibody-antigen interaction. The second assay group that depends on nucleic acid detection is applied for special DNA and RNA sequence detection.

In these assays, the detection sequence is multiplied for DNA by polymerase chain reaction (PCR) method. For RNA detection, RNA is firstly converted to complementary DNA through real-time (RT) method and then follows PCR (RT-PCR) enrichment or nucleic acid sequence-based amplification (NASBA). After multiplication, the enriched molecules (amplicons) are determined according to their size with electrophoresis in agarose gel (they have been previously died with ethidium bromide), and their possible functional identity is certified through hybridization with specialized probe molecules. Finally, the safest procedure is to find primary structure molecules with base sequencing. Nowadays, many specialized analytical assays for substance determination are available but require special equipment, analytical labs, and trained personnel. For this reason, there is a continuous effort to find simple, fast, and low-cost methods to selectively detect the quantity of substances of interest. This determination should be available without the use of expensive equipment and from nonspecialists. The speed has great interest, especially when controlling production in industrial scale. For example, if scientists are able to detect a great concentration of an infectious factor in food, they will be able to stop production and eliminate the damages. In this case, the use of biosensors, devices that convert the biological receptor interaction with the molecule (substance) that needs determination into an analytical signal (optical, electronical, etc.). Chemical sensors are composed of two basic elements: the chemical (molecular) recognition system and the physicochemical converter. The biochemical sensors are chemical sensors where the recognition system is based on a biochemical mechanism. During the past years, there is an intense research activity in the field of liposomes used to develop biosensors. The liposomal size and the number of bilayers can be adapted depending on the production method, so small unilamellar vesicles (SUVs), large unilamellar vesicles (LUVs), or multilamellar vesicles (MLVs) can be created.

Various phospholipid polar heads offer connection ability of various molecules on liposome surface, while the different chain length and their saturation degree allow liposome production of various properties. Additionally, other molecules or lipids can be integrated on lipid bilayer and offer desired abilities to liposomes. For example, the addition of phosphatidylglycerol creates negatively charged liposomes, while the addition of cholesterol lowers the membrane permeability. Due to their great outside surface and their ability to connect to other molecule lipid bilayers, liposomes can be applied in analytical determinations. Liposome advantages include their great inner volume that allows a large amount of dye to be encapsulated (or another appropriate indicator), enhancing the received signal. The analytical determinations are distinguished as homogeneous and heterogeneous. Homogeneous are the determinations where all substances are mixed together in a container and all the reactions take place without any separation step. Heterogeneous determination needs one or more separation stages to achieve excess reagent withdrawal.

The basic factors affecting liposome physicochemical stability are their composition, i.e., the kind of lipids that they are composed of, the number of the layers (production method), their charge, their water-binding ability, and the lipid concentration. Modifying these factors, liposomes that are produced have the desired properties and physicochemical characteristics.

Liposome Physicochemical Characterization and Their Physical Stability

The characterization of liposome physical properties involves techniques and measurements for size, size distribution, surface charge, and ζ-potential determination. Liposome dispersion system stability relates to its physical, chemical, and biological stability. Liposome size and size distribution measurements around the average value are important parameters to calculate the liposome dispersion physical stability. More in detail, we can mention the following: liposomes developed and dispersed in chosen water medium are in constant movement (Brown movement). The particle movement speed depends on their size, solution temperature, and viscosity. Liposome size distribution is a measure of their stability. When size distribution is stable in time, the liposome dispersion is characterized as stable and suitable for pharmaceutical use. The liposome nanosystem stability is time dependent since according to the second thermodynamic law, it is leading to collapse. In case where liposome size distribution increases with time, the liposome dispersion system is characterized as unstable and unsuitable for pharmaceutical use. With dynamic light scattering (DLS) technique (see Chap. 2), the Brown movement calculation is possible through data collected from the light scattering. Brown movement calculation will help to determine diffusion coefficient and, therefore, liposome size and size distribution around the average value. Diffusion coefficient (D) and particle size (characterized from their hydrodynamic radius, R_h) are related mathematically with Stokes-Einstein equation (see Chap. 2).

Dynamic light scattering (DLS) technique measures particle size in relation to particle movement in a liquid, Brown movement. The results being received are the diameter of an equivalent sphere with the same particle diffusion coefficient (liposomes) in the specimen. The hydrodynamic diameter is a little larger than the actual specimen liposome diameter due to solvation phenomena and interactions between particles.

Liposomes belong to colloidal dispersion systems and are characterized as lyotropic liquid crystals (see Chap. 1). These particular liposomal lyotropic states are responsible for the *mesophases* taking place in phase transitions and are related to their thermal stress during phase transitions. Their thermal stress takes place during liposome dispersion system storage or during administration in humans. The thermodynamic parameters that affect and participate in physical stability and, therefore, in pharmaceutical effectiveness of the liposomal product are the following:

- T_m: temperature of basic transition from liquid crystalline state to isotropic fluid.
- $\Delta T_{1/2}$: width of the transition at half peak height. Range in the middle of the peak. This temperature range is related to the cooperativity of system phospholipids or phospholipids and enclosed bioactive molecule.
- ΔH: system enthalpy change.
- $C_{p\,max}$: maximum systems' heat capacity under constant pressure.

The identification and study of the phase transitions of liposomal dispersion systems lipid bilayers allow the control over the thermodynamic parameters mentioned above, in order to rationally design the liposomal system with the most satisfactory physical and thermal stability. The design of liposomal nanosystem and their evaluation in technology level and in vivo behavior are linked with their thermodynamic response. At the same time, thermal analysis and more in particular differential scanning calorimetry (DSC) (see Chap. 2) are valuable tools for liposomes' physical stability prediction and physicochemical property interpretation [51] (Fig. 4.4).

The basic electron microscopy methods used in liposomal physical characteristics study are non-flame atomic spectroscopy (NFAS), transmission electron microscopy (TEM), cryogenic TEM (Cryo-TEM), scanning electron microscopy (SEM), freeze fracture electron microscopy (FFEM), scanning tunneling microscopy (STM), scanning force microscopy (SFM), atomic force microscopy (AFM), lateral force microscopy (LFM), cryogenic atomic force microscopy (Cryo-AFM), near-field scanning optical microscopy (NSOM), and magnetic resonance force microscopy (MRFM).

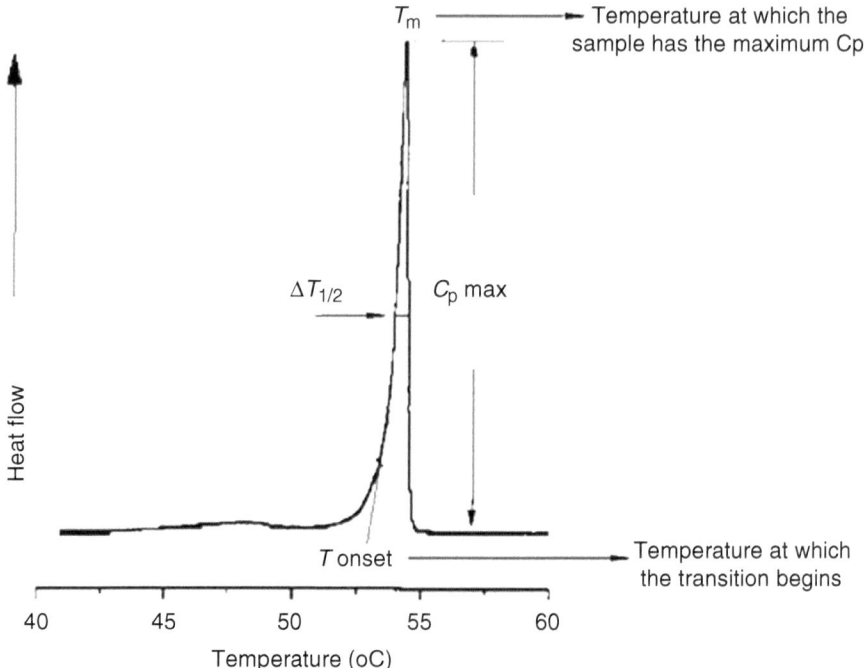

Fig. 4.4 A classical graph of differential scanning calorimetry (DSC) shows the T_m, $\Delta T_{1/2}$, C_{pmax}, and T_{onset} during the phase transitions of the phospholipid dipalmitoyl phosphatidylcholine (DPPC)

Atomic force microscopy (AFM) is an electron microscopy technique (see Chap. 2) used in the study of the structure of liposomal bilayers, their dynamics, and stability. The latter is studied by the use of AFM, and useful calculations concerning the liposomal morphology [57] can be extracted (Fig. 4.5).

Until recently pharmaceutical formulations containing bioactive molecules of anthracyclines, a class of known anticancer antibiotics, have been approved and used in clinical practice. The great interest in anthracycline entrapment in liposomes came up since during their administration they can cause acute and most importantly accumulative cardiotoxicity in patients. To overcome this problem of cardiotoxicity, the following strategies were followed:

- Extended anthracycline administration in order to avoid acute concentration increase in plasma
- Coadministration of substances that prevent free radical formation
- Development of new anthracyclines
- Anthracycline entrapment in liposomes in order to modify their pharmacokinetics and improve their therapeutic index

The formulations already in the market contain doxorubicin in sterically stabilized liposomes (Doxil®/Caelyx®), doxorubicin in conventional liposomes (Myocet) and daunorubicin in conventional liposomes (DaunoXome®) (Table 4.2).

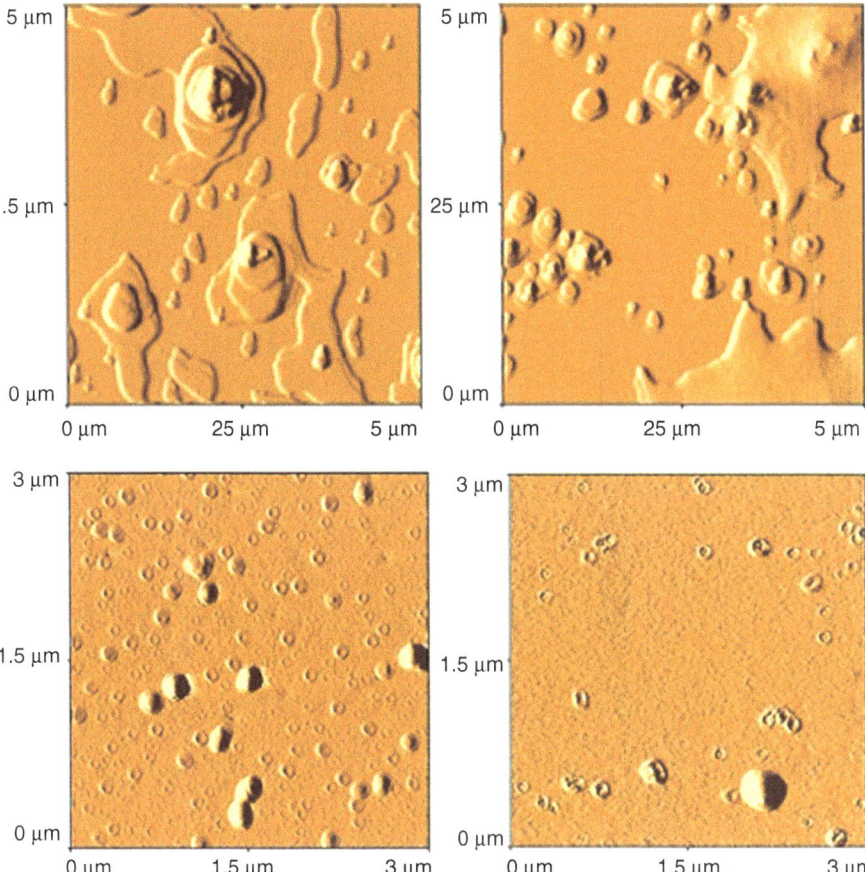

Fig. 4.5 AFM images of lipid bilayers (*top images*) and of liposomes (*down images*) which are composed of egg phosphatidylcholine/dipalmitoyl phosphatidylglycerol (EPC/DPPG) (Adapted from [57]; Cooperation of Laboratory of Pharmaceutical Nanotechnology, Faculty of Pharmacy, University of Athens and the National Technical University of Athens)

These formulations differ in indications, liposome size and lipid composition, and drug release rate. The reason why there are such a small number of liposomal formulations is related to stability and sterility problems as well as final product lifetime. AmBisome® (Gilead Science Inc.) is a liposomal formulation of the polyenic antibiotic amphotericin B. It is a freeze-dried product for intravenous infusion. It consists of hydrogenated soy phosphatidylcholine (HSPC), cholesterol, distearoyl phosphatidylglycerol (DSPG), and α-tocopherol. It is considered as a real liposomal product that consists of unilamellar liposomes with a size of 80 nm. It must also be noted that liposomal product development in large-scale production must follow Good Manufacturing Practices (GMP) as defined from international pharmaceutical control agencies in the United States (Food and Drug administration, FDA) and in Europe (European Medicines Agency, EMA).

Table 4.2 Liposomal medicines in the market

Encapsulated drug	Trade name	Company	Indication	Approval	Innovator company
Amphotericin B	Abelcet	Sigma-Tau PharmaSource, Inc., Indianapolis, IN	Severe fungal infections	1995	The Liposome Company
Amphotericin B	AmBisome	Gilead Sciences, Inc., San Dimas, CA	Severe fungal infections	1997	Vestar
Amphotericin B	Amphotec	Ben Venue Laboratories, Inc., Bedford, OH	Severe fungal infections	1996	Sequus, Pharmaceuticals Inc.
Cytarabine	DepoCyt	Enzon/SkyePharma	Lymphomatous meningitis (intrathecal administration)	1999	Chiron Corporation and SkyePharma
Daunorubicin	DaunoXome	Gilead Sciences, Inc.	Kaposi sarcoma	1996	Gilead
Doxorubicin	Lipodox (generic of Doxil)	TTY Biopharm Company Ltd., Taipei, Taiwan	Kaposi sarcoma, ovarian/breast cancer	2013 (FDA approved; USA)	Sun Pharma
Doxorubicin	Doxil (USA), Caelyx (Europe)	Essex (Europe) Ortho Biotech (USA)	Breast and ovarian cancer, Kaposi sarcoma	1995 (conditional)	Sequus Pharmaceuticals, Inc.
Doxorubicin	Myocet	Novartis Pharma AG, Basel, Switzerland	Breast cancer	2000 (EU)	The Liposome Company
Irinotecan	Onivyde	Merrimack Pharmaceutical Inc. of Cambridge, Massachusetts	Advanced pancreatic cancer	2015 (FDA approved; USA)	Merrimack Pharmaceuticals
Verteporfin	Visudyne	Novartis Pharma AG, Basel, Switzerland	Age-related molecular degeneration, pathologic myopia, ocular histoplasmosis	2000	QLT
Vincristine	Marqibo	Spectrum Pharmaceuticals Inc.	Philadelphia chromosome-negative (Ph−) acute lymphoblastic leukemia (ALL)	2012 (FDA approved; USA)	Inex and Enzon

Quality control during liposome production stages and appraisal and method reliability are extremely difficult and very expensive, especially in the production of medicines where the bioactive molecules are enclosed into liposome membranes. Unfortunately for the liposomal products in the market, industries do not release their full methodology or give incomplete data. Despite all these difficulties, there is a great activity and a lot of interest from pharmaceutical industries to develop new liposomal pharmaceutical formulations since liposomal technology offer solutions in drug administration that presents problems (e.g., paclitaxel) and improves their therapeutic index (Table 4.3). Recently, a liposomal formulation of the anticancer bioactive molecule, i.e., irinotecan, has been approved by the FDA against advanced pancreatic cancer (Onivyde®, Merrimack Pharmaceutical Inc. of Cambridge, Massachusetts) (Table 4.2). So, in the near future, scientists expect the approval of more new liposomal anticancer medications.

Liposomes' most important property – apart from their ability to protect the bioactive molecule from the enzyme degradation, the fact that they have low toxicity without immune response and are biodegradable – is that they can accumulate due to the enhanced permeability and retention effect (EPR effect). This phenomenon is based upon the differences of the tumor and the healthy tissue blood vessel network. So tumor vessels have better permeability since they are developed in greater speed to support tumor's fast development. Apart from this fact, cancer cells are not that thickly

Table 4.3 Liposomal anticancer drugs in clinical phases

Encapsulated drug	Trade name	Company	Indication	Clinical phase
Cisplatin	SPI-077	Sequus	Advanced cancer	I/II
Cisplatin	Lipoplatin	Regulon	Lung cancer	III
Oxaliplatin	Aroplatin	Antigenics	Rectal cancer	II
Vincristine	Marqibo	Inex/Enzon	Non-Hodgkin's lymphomas, acute lymphatic leukemia, Hodgkin's lymphomas (phase II/III) metastatic malignant uveal melanoma (phase III)	II/III
Lurtotecan	OSI0211	OSI	Ovarian cancer, microcellular lung cancer	III
Irinotecan metabolite	SN-38	NeoPharm	Rectal and lung cancer	I/II
Topotecan	INX-0076	Inex	Advanced cancer	I/II
Paclitaxel	LEP ETU	NeoPharm	Breast, ovarian, and lung cancer	I/II
Doxorubicin	ThermoDox	Celsion Corporation, Lawrenceville, NJ	Non-resectable hepatocellular carcinoma	III
Binolerbin	NX-0125	Inex	Advanced cancer	I

placed next to each other, like healthy cells, and the tumor's lymph system that removes substances and nanoparticles (like liposomes) from tissues and organs is insufficiently developed. Therefore, nanoparticles like biological macromolecules or synthetic polymers with a molecular weight greater than 30–40 kDa and liposomes with a diameter up to 600 nm can penetrate tumor blood vessels and accumulate in cancer tissue and not in healthy tissues or organs. The phenomenon is called passive targeting. Liposomal technology has a lot to offer in disease therapy. It must be mentioned that a literature review between 1970 and 2007 related to publications and patents through Scopus TM data (Elsevier B.V) presents 95,082 reports on liposomal technology, 30,979 reports on polymer systems connected with bioactive molecules, and 7453 on copolymers that are used as bioactive molecule delivery systems.

4.1.1.2 Solid Lipid Nanoparticles (SLN)

Solid lipid nanoparticles (SLN) that appeared in bibliography in 1991 are alternative bioactive molecule transfer systems to colloidal nanocarriers, i.e., colloidal dispersions like nanoemulsions, liposomes, and polymeric nanoparticles [58]. These nanoparticles having a size range of 50–1000 nm and the newly categories of nanostructured lipid carriers (NCL) combine more advantages in comparison with classical systems. They can be characterized as safe and effective nanosystems due to biodegradability and biocompatibility. Solid nanoparticles of lipid nature are developed using the homogeneity and microemulsion production method. The lipids of choice are tristearins, stearic acids, cholesterol, and cetyl palmitate. When homogeneity method is applied, the bioactive molecule is dissolved in melted lipids in temperature 5–10 °C over the melting point. The method of thermal homogeneity includes the bioactive molecule dissolution and its dispersion in the melted lipids by continuous mixing in a heated surfactant solution, in the same temperature. The preemulsion that is developed is homogenized to get the nanoemulsion and then left to cool down in room temperature. The parameters that can affect the nanoparticle size and the bioactive molecule entrapment percentage are the following:

- The type of homogeneity
- The homogeneity speed
- The cooling rate in case of thermal homogeneity

The method of cold homogeneity is applied for high-sensitivity hydrophilic bioactive molecules. The lipid particles are dispersed in a cold surfactant solution that is homogenized in a temperature less than room temperature. This process minimizes the lipid melting and, therefore, minimizes the hydrophilic bioactive molecule loss in the water medium.

4.1.1.3 Nanoemulsions

Emulsions are considered as liquid-liquid immiscible dispersion systems. The dispersion phase presents as the low-volume percentage. Emulsions with a dispersion phase particle size of nanoscale are called nanoemulsions [21].

According to the characteristics regarding phase transitions, nanoemulsions can be classified into the following:

- Oil/water (oleum/water, O/W) nanoemulsions
- Water/oil (water/oleum, W/O) nanoemulsions

Nanoemulsions differ not only in dispersion phase particle size but in their properties when related to microemulsions whose particle size is in micrometer (μm).

Nanoemulsion development is not a spontaneous process but requires energy like other nanoparticle categories, i.e., liposomes. So they are characterized as thermodynamically unstable dispersion systems and this relates to their stability. The rational nanoemulsion ingredient choice, the production process, the production temperature, and the raw ingredient concentration, especially the emulsifiers, are critical parameters related to their physicochemical stability and effectiveness.

Nanoemulsions differ from microemulsions since they are transparent (light scattering in a non-visible wavelength) due to dispersed particle nanodimensions and depend on the volume fraction of the dispersion phase. In dimensions greater than nanoscale, dispersion systems are cloudy. An important observation to be mentioned is the surfactant concentration, which in microemulsions is greater than 20 % – nanoemulsions can be produced using surfactant concentrations less than 10 %. The use of nonionic polymers as surfactants is useful in order to avoid and neutralize interaction forces between nanoparticles. It should also be mentioned that the oily phase choice of low viscosity presents advantages of smaller dispersed nanoparticle dimension production in nanoemulsions when compared to oils of high viscosity (triglycerides with long fatty acids).

Generally, nanoemulsion stability studies follow the rules governing the dispersion nanosystem stability studies (see Chap. 2).

The nanoemulsion physicochemical property characterization (size, size distribution, surface charge, ζ-potential, osmolarity, conductivity) involves techniques and observations like the ones presented in the nanosystems mentioned previously (see Chap. 2). For commercial use, long-term stability studies should be performed, especially for their use as drug delivery systems.

Nanoemulsions present advantages in relation to classic emulsions. The most important advantages are the following:

- Due to particle nanoscale, they have a greater outside surface and free energy when compared to common emulsions.
- Do not present the dispersion system's common stability problems.
- Depending on their composition, they can be characterized as nonirritants; therefore, they can be used for skin product development.
- The surfactant choice (that has been approved for human use) allows the parenteral route of administration.

According to all of the advantages mentioned above, nanoemulsions can be used as products for skin care and therefore for cosmetic use. They can be used as lipophilic bioactive molecule carriers due to their lipophilic inside, a fact that makes them more favorable in comparison to liposomes. Their applications are related to their act against bacteria, viruses, fungus, and seeds.

The important antimicrobial action relates to concentrations that do not affect the skin functionality. Last but not least, nanoemulsions can be used in applications like detergents, bioactive molecule delivery systems (parenteral, intravenous, per os, etc.), and intraocular systems.

4.1.2 Polymers

Polymers are substances of high molecular weight, consisted of repeated units called monomers that are connected onto a long chain. Polymer molecules can be linear or branched, while the linear or branched chains can be linked with covalent bonds. Polymers that consisted of the same monomers are called homopolymers. Polymers that consisted of more than one type of monomers are called copolymers [3, 70]. The polymer schematic representation is shown in Fig. 4.6. Polymers do not form perfect crystals but have semicrystalline and amorphous areas (Fig. 4.7).

Various copolymer types can be consisted of monomers with different orientation. Therefore:

- Various monomers can be oriented in linear chain in random or specific alteration along the chain.
- Linear polymer chains can be in monomer block systems called block copolymers. These can be:

  - Diblock copolymers: AB diblocks
  - Triblock copolymers: ABA or BAB triblocks like poloxamers where the chain A is polyethylene or chain B is polyoxypropylene

$$HO\left(CH_2CH_2O\right)_x - \left(CH\left(CH_3\right)CH_2O\right)_y - \left(CHCH_2O\right)_x - H$$

  - where x and y are the monomer numbers in each block.

Fig. 4.6 Structures of homopolymers and copolymers composed monomers A and/or B. (*1* linear homopolymer, *2* alternating copolymer, *3* random copolymer, *4* block copolymer, *5* graft copolymers) (https://en.wikipedia.org/wiki/Copolymer#/media/File:Copolymers.svg)

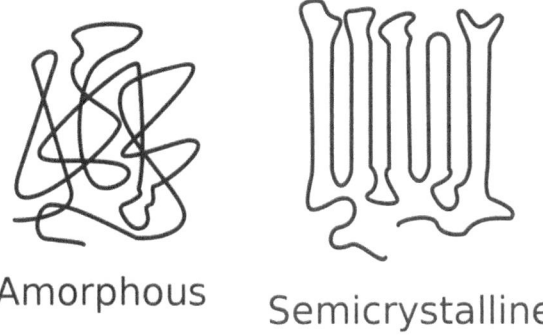

Amorphous   Semicrystalline

Fig. 4.7 Schematic representation of an amorphous and semicrystalline polymer (https://en.wiki-pedia.org/wiki/Crystallization_of_polymers#/media/File:Polymerketten_-_amorph_und_kristallinEN.svg)

The chains can be composed of repeated units from a monomer on which there are grafted chains of a second monomer, like a brush. These polymers are called grafted copolymers [78, 86].

The polymer melting point cannot be absolutely defined, like in low molecular weight crystalline solids due to areas not perfectly structured that are melted in a temperature range. Also, polymer melting point presents variation from the glass transition temperature T_g (g: glass). In a temperature lower than the glass transition temperature (T_g), polymer chains are rigid-immobile and the polymer is glassy and fragile.

In temperatures above the T_g, polymer chains present mobility. Polymers are products that are widely used in the pharmaceutical industry in various pharmacotechnological formulations and in bioactive molecule delivery system coating. Polymer behavior is directly related to their chemical structure. Also, their abilities depend on the way that monomers are connected to each other. Polymers can have linear or branched chains that may cross each other. Copolymers are composed of more than one monomer and develop new polymers with completely new properties. An important polymer application is the genetic material DNA transfer. Concluding the polymer section, important facts for their development must be mentioned:

- Possible toxicity due to cationic polymer use that will develop DNA complexes.
- System physicochemical instability during storage and agglomeration development results in polymer nanoparticle changes in size and size distribution.
- Minimization of cell target transfection ability.
- System stability problems while in the organism.
- Possible final product high-cost production in industrial scale (scale-up).

A particular class of polymers is the class of polyelectrolytes. A polyelectrolyte, according to IUPAC, is considered to be a macromolecule that has ionic groups or groups that are able to ionize [35]. In other words, a polyelectrolyte is a polymer that consists of one repeating ionizable group along their backbone. This group can dissociate in a polar solvent such as water and leave counterions resulting in an opposite charge on the backbone. Polyelectrolyte block copolymers constitute an intriguing class of bio-inspired macromolecules, as they combine the structural properties of amphiphilic block copolymers, polyelectrolytes, and surfactants and provide various possibilities for use as nanostructural delivery platforms of genes and proteins/peptides, through electrostatic complexation of the pharmaceutical agent. The polyelectrolytes' solution shows particular properties and behavior that are different from those exhibited in neutral polymer solutions and solutions of electrolytes [69]. They are water soluble and are considered to be promising drug delivery systems especially therapeutic biomolecules like protein and peptides (see Chap. 5) [10]. They are also used as stabilizing agents in colloidal dispersions and as rheology modifiers as well as in pharmaceutics, as suspending agents [11].

It is important to mention that DNA and RNA are polyanions due to their negatively charged phosphate esters in their backbone [8] and are belonging to the biological polyelectrolytes. Several other biological molecules such as proteins and polysaccharides are charged macromolecules with essential functions for the living organisms. Association of biological polyelectrolytes with synthetic polyelectrolytes or polyelectrolyte complexes is of great importance in drug delivery and in biopharmaceuticals [49]. The application of polyion complex micelles into therapeutic fields is rapidly increasing due to the simple and efficient encapsulation of biopharmaceuticals like insulin, lysozyme, antitumor peptides, and DNA/RNA and outstanding biocompatibility among various polymer-based nanocarrier delivery platforms. Proteins and polyelectrolytes interact, primarily via electrostatic interactions, to form hybrid complexes, which can have widely varied stoichiometries, morphologies, architectures, and shapes. These mixed systems are bio-functional with potential applications in the design and development of delivery systems of biopharmaceuticals.

Polylysine-based block copolymers are suitable nanocarriers to transport and deliver proteins, peptides, and genes. These mixed biomaterials exhibit both pH- and temperature-responsive behaviors and self-assembly properties and are used as nanocarriers with enhanced properties. According to Lee and Kataoka [43], ionic biopharmaceuticals, such as genes and proteins, can interact with ionic block copolymers to form polyion complex micelles with core-shell morphology. Insulin is an attractive biomolecule to encapsulate into polyion complex micelles. It is a polyampholyte with isoelectric point at pH 5.5 and can be either positively or negatively charged due to both basic and acidic groups. The encapsulation of insulin into nanoparticles is a promising strategy that has been developed in order to enhance its absorption and its bioavailability, aiming a successful delivery. Several approaches have been employed in order to realize effective insulin formulations. The structural analysis of insulin in pharmaceutical formulations recently appeared in the literature. An ideal insulin carrier should have reasonably high protein encapsulation efficiency and loading capacity and sustained/controlled release of the loaded protein while retaining bioactivity [7, 30, 79].

4.1.2.1   Polymersomes

The evolutions in the field of controlled polymerization techniques have allowed the design of a new category of amphiphilic membranes composed of block copolymers.

Block copolymers are macromolecules that have one or more polymer types as mentioned earlier. This combination of various polymers resulted in new properties. Polymer amphiphilic bonds have the ability of self-assembly into complex membranes offering stability and improved mechanical properties in comparison to the conventional phospholipid membranes. The simplest structure that these membranes can form is a *pseudo-spherical* shell known as polymersome [84]. Polymersomes are interesting structures that can be self-assembled and can be used for bioactive molecule delivery to the damaged tissues. Their ability to encapsulate hydrophilic molecules and integrate hydrophobic bioactive molecules makes them attractive delivery systems. But they demand specific physicochemical parameter values for their development and a specific ratio of hydrophilic/hydrophobic section for, i.e., a block copolymer to form vesicular structures.

Biophysically speaking, polymersomes – and liposomes as mentioned before – are interesting biological membrane simulation systems (membrane mimics), while intermembrane proteins can be integrated with reconstitution methods leading to proteopolymersomes [59]. On the other side, amphiphilic block copolymers have been widely studied for biomedical and pharmaceutical applications from the controlled, sustained, delayed-release advanced technology to membrane-mimic and bioactive molecule transport.

Polymersomes with dimensions varying from 10 nm to 10 μm have a relatively high control on size distribution around the average value of the dispersion medium. Polymersomes have drowned scientists' attention to study innovative nanosystems for various applications in different scientific fields. More in specific in the field of pharmaceutics, polymersomes have the ability of incorporating a wide range of bioactive molecules and biomolecules and offer control over their release rate. Amphiphilic molecule (like phospholipids) self-assembly with various chemical structures is related to their geometrical characteristics (Thompson 2012). The geometry of these structures is defined from the proportion of hydrophobic-hydrophilic section of the amphiphilic molecule. This self-assembly amphiphilic molecule approach through geometric characterization has facilitated the understanding not only of phospholipids but of low molecular weight surfactants as well. The high molecular weight molecules like amphiphilic block copolymers, polypeptides. etc. can be designed to have the same amphiphilic character as phospholipids and surfactants but also composed of polymer chains covalently linked in series of two or more blocks. It is known that block copolymers can be organized in a great structure range, for example, in lamellar structures. Lamellar structure hydration with water solution results in a steady dispersion of amphiphile block copolymers. Proportional to the packing parameter S_p (see Chap. 4), the factor F corresponds to the hydrophilic fraction and determines the morphology of copolymers formed in aqueous media. In the case of polymersomes (unilamellar polymer vesicles), F is between 25 and 40%.

Initial copolymer studies with F values that form polymersomes having a molecular weight (MW) between 2000 and ~20,000 Da have shown through cryogenic transmission electron microscopy (cryo-TEM) that the membrane core (d) increases by molecular weight increase from d ~8–21 nm. Simulations that took place revealed that, only in small molecular weight systems, the polymorphic bilayers show a clear average level of high density that strongly reminds "methyl through" like the one appearing in lipid bilayers. For the high molecular weight copolymers, the two layers of the lipid bilayer are mixed or melted together in a homogenous thick shell. It is possible to produce diblock copolymers, triblock copolymers, multiblock copolymers, and grafted copolymers by controlling their synthesis. This observed chemical architecture variety can be of use when designing various membrane polymersomes with various structures. Polymersomes produced from AB diblock copolymers present interconnected membrane. The powerful entanglement inside the hydrophobic layer can be granted as physical crossed connection enhancing their mechanical strength in comparison to conventional liposomes.

Triblock copolymers BAB, i.e., hydrophobic-hydrophilic-hydrophobic copolymers, are similar with diblock copolymers since there is only one molecular structure that leads to membrane development: hydrophobic parts of the chain are put together to form the membrane and the hydrophilic diblock forms a U-shaped loop. BAB diblock copolymers cannot form a loop where the two hydrophobic diblocks link at the same part of the hydrophilic diblock. The membrane structure can be additionally evolved by using multiblocks that have various polymers and are subjected into phase separation after assemblage. When the proportion of hydrophobic/hydrophilic section favors the membrane development, ABC copolymers are formed into asymmetric membranes that form polymersomes whose inner and outer chemistry is different in order to minimize the surface tension and enhance the vesicle curvature (Fig. 4.8). Polymersomes, when making block copolymers, assemble in lyotropic phases of the liquid crystalline phase, like reverse hexagonal structures, lamellar, hexagonal perforated membranes, *sponge phases*, etc. Figure 4.9 shows a detailed representation where the formed structure is presented as a proportion to the polymersome molecular weight and concentration in water. The polymersome size significantly affects the lyotropic phase development and the dispersed vesicle structures. The evolution from solid to lyotropic liquid crystal shows that small molecular weight copolymers are formed initially in reversible hexagonal structure and then in lamellar, while the high molecular weight copolymers are dissolved directly to lamellar structure. The transition from lamellar to *sponge phase* is not affected by the amphiphilic molecule molecular weight; instead the transition from sponginess structure to vesicles is quantitatively different and depends on the amphiphilic diblock copolymer size. Smaller copolymers form dispersed vesicles, while high molecular weight copolymers initially form vesicle gel bluster diblocks that are finally disrupted in dispersed vesicles. Amphiphilic molecules' molecular weight affects the nature of the developed vesicles. Large amphiphilic molecules form exclusively unilamellar vesicles and smaller copolymers form multilamellar vesicles. Copolymers' molecular weight does not only affect phase transition but the kinetic transition from phase to phase. One of the most important diblock copolymer clusters is their non-ergodic nature, for their formation in organized structures. This practically means that there is no structural unit exchange (polymeric chains) between micelles/vesicles and solution, underlying a critical agglomeration concentration near zero. This is a desired property for polymersomes and causes long lifetime.

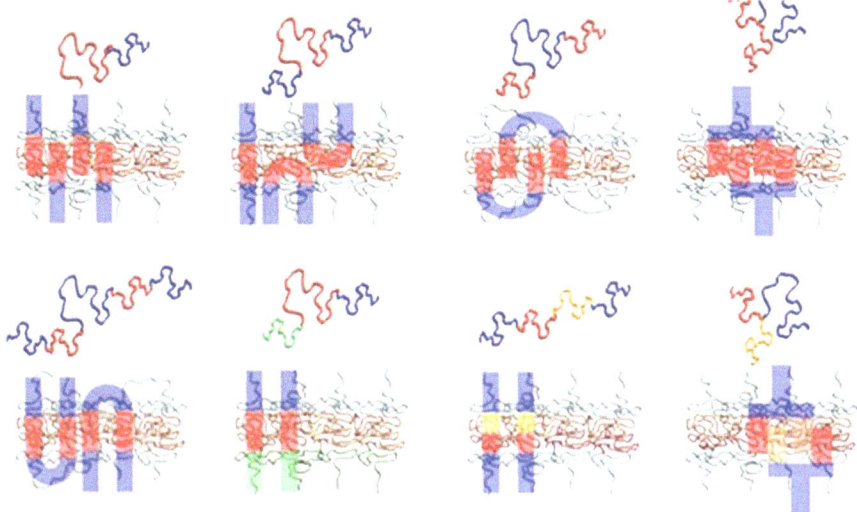

Fig. 4.8 Membrane structure of polymersomes formed by diblock (AB), triblock (ABA, BAB, ABC), multiblock copolymers and micto-arm copolymers (Adapted from [50] with permission from Springer)

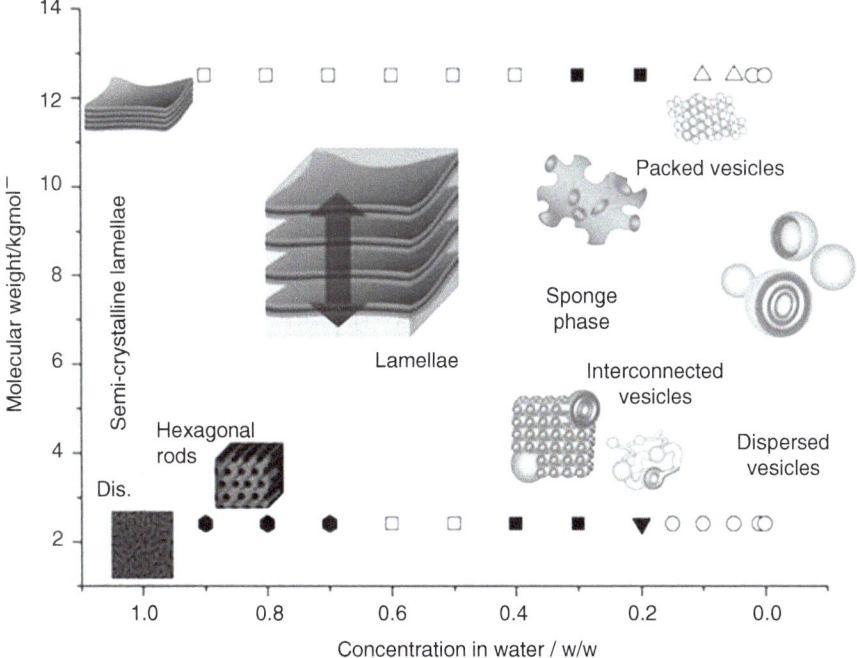

Fig. 4.9 Phase diagram of copolymer in water (Adapted from [50] with permission from Springer)

Polymersomes that form diblock copolymers swell when in contact with water in two different phases. Initially, the water and copolymer are diffused into each another resulting in amphiphilic membrane molecular setup. After a critical time where the polymer's molecular weight is exponentially changed, the amphiphilic membranes will gain balanced structure, and therefore, the formation will follow Fick's diffusion. This kind of complex movement shows why polymersomes were studied later than copolymer micelles. Polymersome production method is presented in the Appendix.

Polymersome Size and Size Distribution Appraisal

Polymersome size and size distribution are critical parameters to study in order to use polymersomes as bioactive molecule carrier and delivery nanosystems. Polymersome size and size distribution appraisal define in vitro and in vivo nanoparticle effectiveness. In vivo, the size defines the nanosystem circulation times and their ability to reach specific targets, the extravasation possibility, their properties, and their final decomposition and clearance. It is not yet completely understood which parameters affect polymersome size and size distribution.

Regarding polymersome formation, it was observed that the proportion of hydrophobic/hydrophilic copolymer section seems like micelle structure (i.e., the copolymer is mostly hydrophilic), the mean vesicle diameter decreases to three times down. This is attributed to the curve formed and it is obvious that the more hydrophilic the copolymer is, the more curved and stable the polymer structures are. On the contrary, crystalline development can show that the polymersome size is reversely proportional to the copolymer concentration. Under mixing conditions, where the concentration is high, polymersomes appear to be in nanometer-size diameter, while in mild mixing conditions, their diameter is in μm size. Experimentally, polymersome size is strictly defined from the method used for their development (Fig. 4.10).

Polymersome Properties and Applications

Polymersomes have the ability to integrate into their membrane an important amount of hydrophobic bioactive molecules. This integration prevents self-aggregation that would normally take place when these molecules were in solution in their free form and protects them from interactions with biological components.

The polymersome amphiphilic nature allows the intake and integration of amphiphilic molecules on their membrane. Polymersome macromolecular nature makes them especially resistant in detergent dissolution, and therefore, the surfactant integration is proposed to be on the polymersome membrane to avoid collapse. Ruysschaert and coworkers [71] observed that this hybrid structure helps implementing phospholipids inside polymersomes, a fact that offers them advantages related to their physicochemical properties and greater effectiveness as bioactive molecule transports. Possibly, the most important example of amphiphilic molecule stabilization is the use of polymersomes for cell membrane protein scaffold. The ability of hydrophilic molecule entrapment into polymersomes is an important

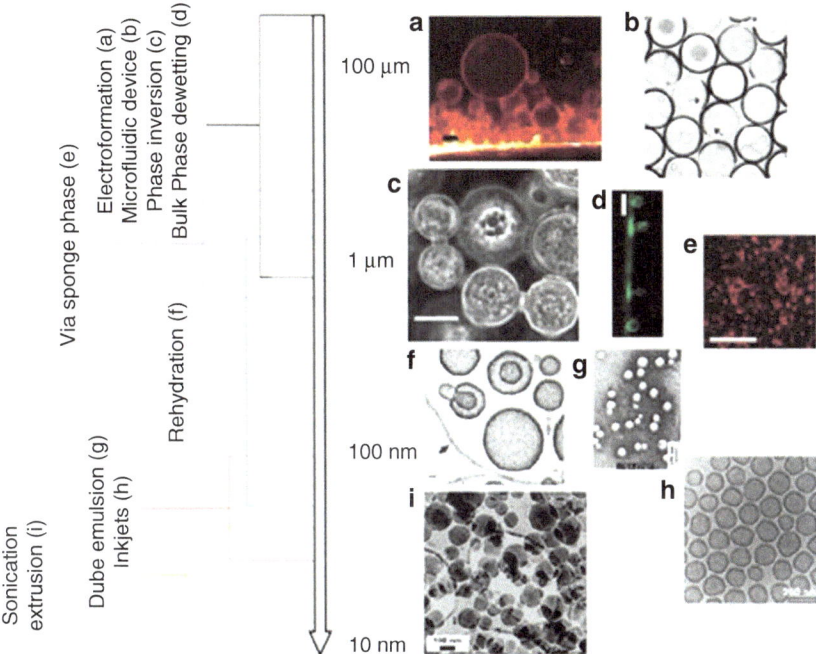

Fig. 4.10 Schematic representation of the size distribution of polymersomes, which depends on their manufacture methodology (Adapted from [50] with permission from Springer)

advantage for their transport and delivery to damaged tissues. Despite all the above, the hydrophobic or amphiphilic molecule entrapment is more or less simple; the complex kinetics when forming polymersomes prevent hydrophilic molecule entrapment. The most important parameter regarding hydrophilic molecule entrapment is the polymer membrane permeability. One of the advantages that polymersomes have is that polymer solution properties are strictly checked according to side groups' chemistry. The same polymer chain has both soluble and insoluble parts and their equilibrium defines total solubility. Therefore, polymer solubility can change from outer stimulants like temperature, pH, ionic strength, light, etc. This polymersome response to the outside environment has been used in polymer mechanics, and many devices have been developed based on polymer solubility. In polymersomes, these properties can be used for structure development that will be dissolved under specific environmental conditions. The simplest approach has been made with hydrophilic polymer connection with hydrolyzed polymers. In contact with water, polymers will be degraded by nonreversible polymersome disassembly and content release. The degradation ability after a period of time or an external stimulant is of critical importance for new carriers and for the modified polymersome diameter development. Reversible disassembly can take place when combining hydrophilic polymers with polymers presenting pH, temperature, and radiation-dependent solubility. For example, polymersomes with structural units of poly(L-glutamic acid)-poly(L-lysine) that respond to pH can be modified reversibly into weak acidic or basic water solutions. Lastly, similar transition can take place by theuseofsensitivetoradiationgroupslikepoly(ethyleneoxide)-poly(methylphenylsilane) (PEO-PMPS) and azobenzene-containing poly(methacrylate)-poly(acrylic acid) (PAA-PAzoMA). Recently, Mabrouk and his coworkers [48] published structures of asymmetrical polymersomes whose membrane consisted only of one lamellar copolymer sensitive to radiation: poly(ethyleneglycol)-poly(4butyloxy-2-(4-(methacryloyloxy)butyloxy)-4 (4-butyloxybenzoyloxy)azobenzene (PEG-b-PMAazo444). When exposed to radiation, these polymersomes burst. This fact is attributed to thermodynamic changes related to membrane curve.

Polymersomes' Surface Chemistry

Polymersome membrane is the result of amphiphilic diblock copolymer auto-assembly in water, as mentioned earlier. Also chemical structure can be used offering the necessary hydrophilic/hydrophobic conditions to preserve the assembly. As mentioned before, hydrophilic blocks concentrate in brush configurations, which according to polymer nature will modify-control the polymersome surface characteristics and its interaction with the environment. As an example, it can be mentioned that block copolymers have polyethylene oxide (PEO) that assemble in polymersomes with highly hydrated and neutral polymeric brush orientation, having little protein interactions. This orientation allows PEO polymersome structural integrity in biological fluids without immune systems interactions.

Their ability to escape the immune system is known as "stealth" and has been widely used to increase liposomes and other nanoparticles lifetime in the organism. The brush thickness and structure are critical parameters for polymersome stealth ability, as Photos and coworkers [67] have proven since they found different circulation times for different molecular weight PEO polymersomes. Polymers of the polyethylene glycol (PEG) category and other non-fouling polymers are acceptable for biological applications. It has been recently proven that polymersome surface can be additionally modified with molecules characterized as islands and have different chemistry. This is made possible by mixing various copolymers developed to produce polymersomes that separate phases of polymer-polymer. In a second level of complexity, polymersome surface can withhold macromolecular structure biomaterials like proteins, antibodies, vitamins, hydrocarbons, etc. Biotin groups' connection with hydrophilic diblocks is used to group together abidines to predefined polymersomes that will later on connect to biotinyl targeting ligand (since each abidine group has four connection spots for biotin). Using a similar approach, polymersomes with anti-ICAM-1 antibodies have been tested for inflamed endothelial cell treatment.

Polymersome Applications

Some important polymersome applications concern the in vivo tissue imaging by using near infrared (NIR) and protein, DNA, and bioactive anticancer molecule entrapment on their inside. Apart from an exceptional delivery and a bioactive molecule release system, polymersomes seem to be equally useful in imaging applications.

Additionally, polymersomes offer the ability of hydrophobic and hydrophilic bioactive molecule incorporation, addition of targeting macromolecules on their outer surface, as mentioned and analyzed earlier on, forming the nanocarrier for damaged tissue targeting according to "combined drug delivery" and magnetic resonance imaging (MRI). Another polymersome application is their use as protein incorporation systems, as mentioned earlier. Recent studies on this field concern primary human antibody delivery using fluorine-labeled antibodies. Gene therapy is another important research field mostly due to the polymersome ability to replace or exclude specific gene expression. Intracellular genetic material delivery lacks effectiveness due to their negative charge and size since repulsive interactions exist with the negatively charged plasma cell membrane. Many methods have been explored to avoid interaction with cell membrane by using gene transport carriers inside the cell. Some of these methods include polymersome use [59].

4.1.2.2 Biodegradable Polymeric Nanoparticles

Polymeric nanoparticles are solid, colloidal particles of 10–10,000 nm size. The bioactive molecule can be entrapped inside the nanoparticle, absorbed, or connected on its surface. According to the production method used, polymer nanoparticles, nanospheres, and nanocapsules have different properties and characteristics that affect bioactive molecule release. Nanocapsules are systems where the bioactive molecule is on the inside surrounded by a polymer membrane. Polymeric nanospheres consisted of a matrix where the bioactive molecule is dispersed. The advantages of using polymeric nanoparticles for bioactive molecule administration and transfer are due to their basic properties: small size to achieve small blood vessel penetration, taken over by cells and accumulate in target areas. The use of biodegradable materials allows bioactive molecule steady release rate in the target tissue for a period of time equals days or even weeks.

4.1.2.3 Polymeric Micelles

Amphiphilic copolymers in water solutions can self-assemble, like previously mentioned and form micelles (Fig. 4.11). Polymer micelles have nanometer size, with a hydrophobic core (where lipophilic molecules can be attached) and hydrophilic parts that help in the steady dispersion development in water. Their distribution in the organism depends on its surface properties. Polymer micelles due to their small size (<100 nm) are not recognized from the MPS and can be accumulated in the damaged tissue environment with passive diffusion. Surface macromolecules, like antibodies, can connect on polymer micelles and produce immunomicelles that will present selectivity. Polymeric micelles present advantages over conventional micelles that are

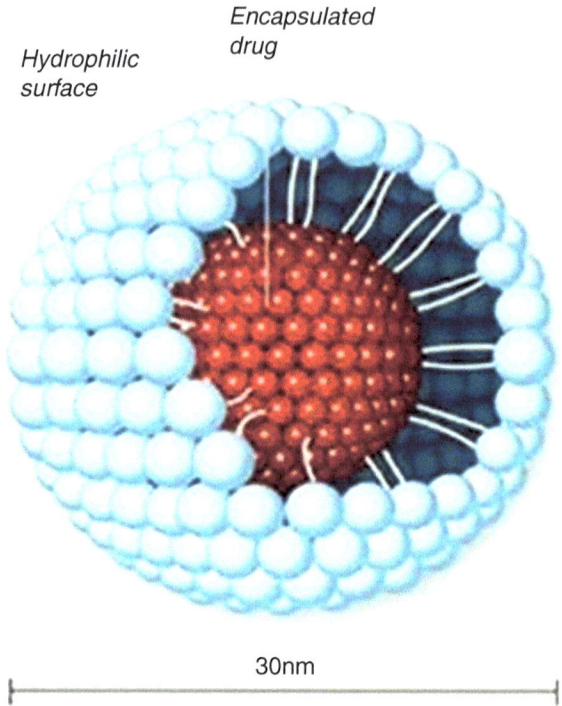

Fig. 4.11 Polymeric micelles (Adapted from www.atrp.gatech.edu/pt18-3/18-3_p3 with modifications)

formed from surfactants because they have better thermodynamic stability in biological solutions, resulting in their slow in vitro disintegration [86].

4.1.2.4   Nanogels

Nanogels are swollen nanosized networks consisted of hydrophilic or amphiphilic polymer chains. Nanogels can protect and transfer bioactive molecules and therapeutic nucleotides and control their release by integrating highly familiar functional groups that correspond in configuration and biodegradable links in the polymer network. Like other nanosystems, nanogels can be easily administered as liquid pharmacotechnological formulations for parenteral administration. Nanoparticle size offers high special surface that is available for bioconjugation with factors targeting special macromolecular targets on damaged tissue surface.

4.1.2.5   Dendrimers

Dendrimers belong to polymers of the fourth generation that were firstly formed in the beginning of the 1980s and are consisted of long chains connected to a central tube.

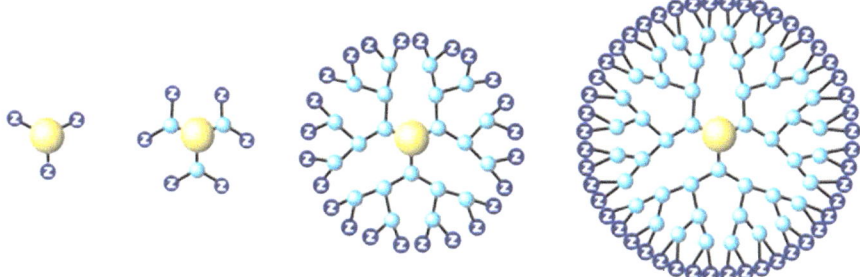

Fig. 4.12 Development of generations of dendrimers (http://www.chemheritage.org/discover/online-resources/chemistry-in-history/themes/microelectronics-and-nanotechnology/tomalia.aspx)

They are structured around a central core and are composed of repeated structural units (Figs. 4.12 and 4.13). They are attractive bioactive molecule transfer nanosystems due to their small size (<10 nm) and the multiple functional groups that can be attached on their surface. Dendrimer interaction with biological environment is defined from their end groups. Dendrimers have special properties due to their shape that looks like a tree and their inner cavities. Bioactive molecules can be enclosed in dendrimer cavities or connect to final surface groups [17, 18, 36]. The term dendrimer comes from the Greek word δένδρο (*dendron* (=tree) and μέρος *meros* (=part) and describes a new group of branched macromolecules whose architecture looks like a tree. In 1985 Donald Tomalia and his coworkers [80] published the composition and the full characterization of a new macromolecular group of the poly(amidoamines) [Poly(amidoamine)(PAMAM)] calling them dendrimers. Until this day, the mostly studied dendrimer group is PAMAM generation that is the most commercially used for research applications. Many new dendrimers are composed of various structural units. Dendrimers have an important research interest as these branched polymers of nanometer size offer conventional methodology of bioactive molecule transfer and delivery, i.e., their structural unit chemical control, their molecular weight, their surface characteristics, and biological targeting process control [81–83].

In comparison to conventional linear polymers, dendrimers have specific dimensions, almost spherical shape and molecular weight and not molecular size distribution. In contrast with other polymers, the critical parameters in nanoscale, like size, shape, and final active groups, can be accurately controlled through their architecture, i.e., core, internal branching units, and final surface groups. These properties are not the sum of the monomer properties but are completely different and follow different rules. The term dendrimer phenomenon is used to describe these unexpected dendrimer properties. Dendrimers, as mentioned earlier, have branched units with end chemical groups that have various properties, allowing charge, hydrophilicity, or hydrophobicity alteration. Dendrimer size is extremely small, 2–10 nm, and therefore can be characterized as true nanoparticles. Their dimensions cannot be extremely large because of their branching unit stereochemistry. Branched polymers presented in the 1960s were the precursors of dendrimers since their branched units are simple branched polymers and can be characterized from their molecular

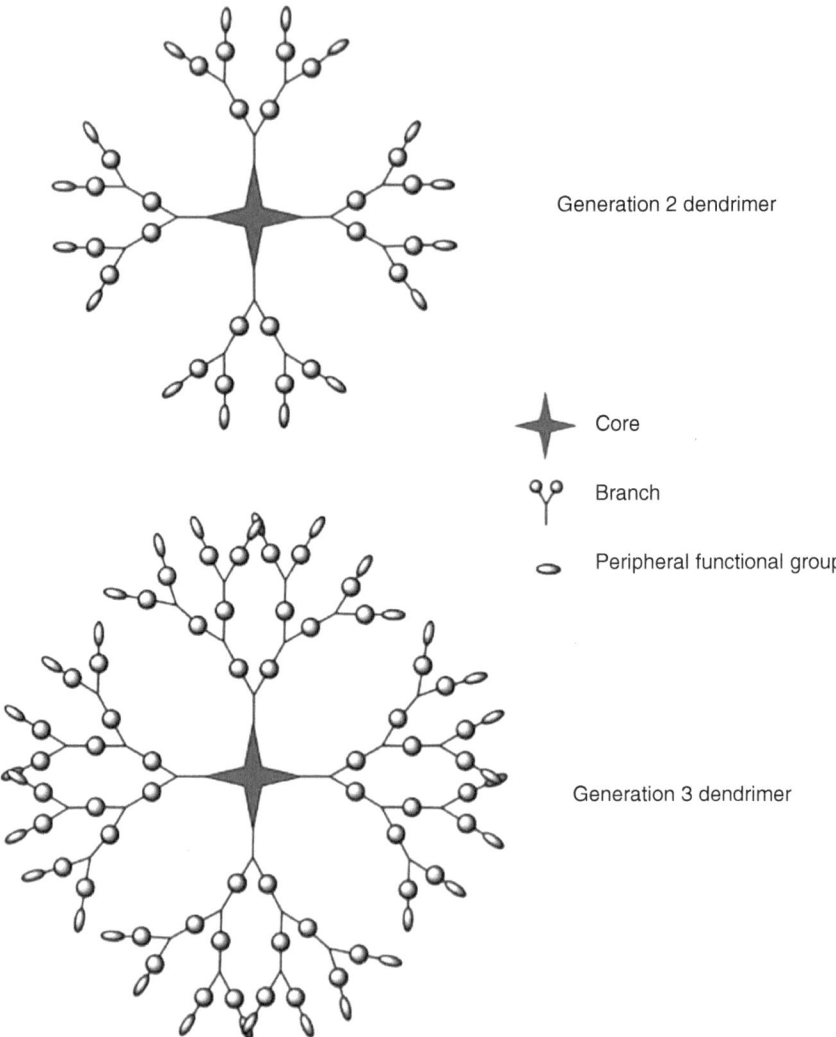

Generation 2 dendrimer

Core

Branch

Peripheral functional groups

Generation 3 dendrimer

Fig. 4.13 Dendrimer structures (generations 2 and 3) which present the cores, the branches, and the peripheral functional groups (Adapted from [24] with permission from Bentham Science Publishers)

weight, structure, and physicochemical characteristics according to technology evolution and analytical assays developed in the past decade. Biologically speaking, dendrimers seem to present architectural similarities with macromolecules like proteins, polysaccharides, and nucleic acids.

Dendrimers are composed of three distinct areas:

- The core where all the branching units start from
- The branching units that compose the dendrimer
- The outer surface where the active groups are placed

The core and the repeated units define the microenvironment inside the molecules, while the surface groups define the solubility and the physical and chemical interactions with the environment. Therefore, the combination of these three areas defines the physical and chemical properties of the dendrimer molecule. Since the innovative works of Tomalia [80] and Newkome [61] which were published independently, the research in this field created intense interest and concentrated on dendrimer design and development with specific physical and chemical properties, aiming on their application on fields of great scientific and commercial interest such as biomedicine.

Dendrimer Use in Biomedicine

There are a large number of applications of dendrimeric structures in biomedicine with emphasis to drug delivery and imaging. However, dendrimers are used for diagnostic and therapeutic purposes like:

- In diagnostics – Gd^{III} block dendrimers are used in magnetic resonance imaging (MRI).
- In DNA biosensors.
- In therapeutics as controlled release drug delivery.
- In gene therapy (gene transfection).
- In cancer therapy with ^{10}B (boron neutron capture therapy).
- As antimicrobial and antivirus medication.

The advantage of dendrimer use as bioactive molecule carriers is based upon their structure control which is a result of their controlled production. This controlled synthesis allows molecule production with special properties, like hydrophilic groups on the outer surface and internal hydrophobic cavities that offer the possibility of different chemical molecules' entrapment. The dendrimer family that has been studied the most as a bioactive molecule carrier is the PAMAM.

Tomalia [81] and other scientific groups have studied the relation between structure and biocompatibility of PAMAM derivatives. They observed that cationic dendrimers (that have NH_2 groups on their surface) are generally toxic and cause hemolysis even in low concentrations. Also, they observed that toxicity increases in greater generations, i.e., seventh generation is much more toxic than fifth or third generation. In contrary, anionic dendrimers (that have COOH groups on their surface) do not present toxicity. Despite these, negatively charged carboxyl dendrimers are not appropriate for delivery systems as they cannot interact or connect with the negatively charged surface. To overcome this problem, PAMAM derivatives have been developed that had hydroxyl groups on their surface, or polyethylene glycols (PEG) chains were connected to outer groups and used as anticancer bioactive molecule entrapment.

Another dendrimer class that has been studied and developed is poly(ester) dendrimers, firstly produced by Frechet [23, 34]. Poly(ester) dendrimers that are usually asymmetric can enclose anticancer molecules like doxorubicin. Also, the development of a liposomal nanosystem and dendrimer (PAMAM) has been tested for the entrapment of a large quantity of the anticancer agent methotrexate and doxorubicin. Two different ways of using dendrimers as bioactive molecule carriers have been described: the bioactive molecule entrapment inside the dendrimer (inside the cavities formed from the branches) through hydrophobic interactions and ionic links and the covalent bonding of the bioactive molecule on the dendrimer surface. The development of a multifunctional dendrimer consisted of PAMAM G5 generation connected with the anticancer agent methotrexate or paclitaxel, folic acid, and fluorescein as detection molecules for dendrimer course into the organism is a contribution to nanosystem development that has targeting, imaging, and therapeutic functions against cancer at the same time. It is well established that cancer cells due to their quick multiplication need large quantities of folic acid. They achieve that by increasing the folic acid receptors on their surface. So, by taking on folic acid that is connected to the dendrimer, anticancer cells receive at the same time the bioactive molecule. Lastly, the fluorescent molecule allows dendrimer tracking. The first in vitro experimental results in cancer cells that overexpress folic acid receptors are encouraging. Some of the bioactive molecules enclosed in or connected with dendrimers are indomethacin, methotrexate, doxorubicin, fluorouracil, paclitaxel, and ibuprofen. Dendrimer synthesis is analyzed in the Appendix of this chapter.

Dendrimer Applications

There are more than 50 dendrimer families, each one with unique properties like surface, internal cavities, and core that can be adjusted in various applications. Polymer multiple possible applications are based on their molecular uniformity, multifunctional surface, and internal cavities. All the above properties make dendrimers appropriate for high technology applications in the fields of biomedicine and industry. More in detail:

• Dendrimers have been used in in vitro diagnostics.
• Dendrimers have been used in preclinical studies.
• There are research efforts for dendrimer use in the delivery of bioactive molecule target tissues. Bioactive molecules can be equally incorporated inside dendrimers or connect to outer groups.
• Dendrimers can be used in industrial processes' improvements.
• Dendrimers have been used as carriers known as "vectors" in gene therapy.

Dendrimers belong to a chemical group that presents special characteristics resulting in a great variety of applications. They belong to nanotechnological systems, and their basic characteristic is the lack of polydispersity due to absolute controlled synthetic reaction process that leads to polymer systems of specific molecular weight. Their almost spherical outer surface allows many outside groups with various functions (targeting, detection, delivery of bioactive molecule), while cavities formed inside can enclose bioactive molecules. The application of dendrimers has been studied in several fields of modern technology. It is important to figure out that the profile of a candidate drug is affected by its solubility in aqueous media, its lipophilicity, degree of ionization, pharmacokinetics and pharmacodynamics, permeability, and its behavior regarding the binding process with plasma proteins. As dendrimers belong to the class of polymers, polymer therapeutics is a term that describes nanosized polymeric carriers that are in clinical phases and practice for diseases such as multiple sclerosis, renal failure, and virucide vaginal formulations [87].

Dendrimers have been used for most hydrophobic anticancer molecule dissolution due to their ability to produce a great amount of hydrogen bonds in water environment. In relation to other nanotechnological systems, liposomes and polymers have been used in order to take advantage of the EPR phenomenon (see Chap. 6). This in vivo anticancer action and the simultaneous toxicity reduction of the known chemotherapeutic agents like doxorubicin and cisplatin when incorporated in dendrimers were attributed to the enhanced permeability and retention (EPR) phenomenon. Poly(glycerol-succinic acid) dendrimers have been used to dissolve 10-hydroxycamptothecin. The system effectiveness was increased in human breast cancer series, but the bioactive molecule quick release (6 h) excluded its systemic administration. Important entrapment percentages in various dendrimer structures have been achieved for anticancer molecule etoposide, paclitaxel, methotrexate, and 6-mercaptopurine. The hydrophobic anticancer molecule entrapment in middle size dendrimers (G4–G6 generation) has been proven to increase their solubility and toxicity. The most basic disadvantage is the lack of kinetic controlled release since these systems release their content in a few hours after systemic administration. Possibly, dendrimer systems with incorporated anticancer agents can be used in the future for infusion directly to the tumor. The method of bioactive molecule covalent bonding on dendrimer surface offers advantages in comparison to their cavities. Different bioactive molecules can be connected to a dendrimer molecule and their release depends in the linkers' nature. With this method, methotrexate, paclitaxel, doxorubicin, and 5-fluorouracil have been covalently bonded with dendrimers of the polyamidoamine group (PAMAM). In all occasions, the results showed effectiveness sustenance, delayed release, selective accumulation on cancer cell, and lower toxicity. Dendrimers can be applied as viruses' inhibitors, and their activities has been shown against HSV, RSV, and HIV, while a dendrimeric formulation as vaginal gel (i.e., VivaGel™, Starpharma) is in clinical trials.

Active Targeting

Taking full advantage of dendrimer architecture allows simultaneous bioactive molecule and target group connection on the same dendrimer molecule. In the field of oncology, targeted transport of chemotherapeutic molecules in cancer cells is translated as side effect reduction since healthy tissues like the liver, spleen, and bone marrow can accumulate bioactive molecule toxic levels. For dendrimer system active targeting, monoclonal antibodies, PEG chains, and imaging agents with the encapsulated drug have been used for the production of multifunctional dendrimers

Fig. 4.14 Schematic
representation of
multifunctional dendrimers
carrying on their surface
antibodies, PEG, and
imaging agents with
encapsulated drug

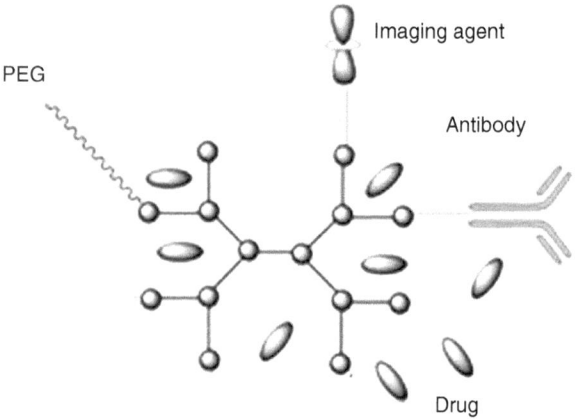

(Fig. 4.14). Dendrimer great surface in relation to their size and high solubility makes them useful as catalysts since they can combine the advantages of homogenous and heterogeneous catalysts.

Dendrimer Application in Imaging with Magnetic Resonance Imaging Technique

An important diagnostic method is the magnetic resonance imaging (MRI) that is the spectroscopy application of the nuclear magnetic resonance (NMR) that offers three-dimensional imaging of the living organs and their blood vessels. The use of paramagnetic metal cations (enhancer parameters) improves the method sensitivity and accuracy. A very common enhancer parameter, the gadolinium salt diethylene triamine pentaacetatic acid (DTPA), is unfortunately diffused outside the blood vessels due to its low molecular mass. To overcome this problem, scientists studied dendrimers with gadolinium salts on their outer region. These studies showed that the modifications of dendrimer characteristics, like size and nuclear or regional composition, can lead to the development of enhancer factors that will be specialized for specific organ imaging or lymph node blood vessel imaging. Additionally, the use of dendrimers makes targeted delivery possible since the dendrimer molecular structure can be transferred to cancer tissues making dendrimers useful tools in cancer imaging. An example of dendrimer application in cancer treatment is in boron neuron capture therapy (BNCT), an experimental method of two steps. Initially, the patient is injected with a boron 10B radioactive isotope which selectively accumulates in cancer cells. Following, the patient receives radiation of a low-energy neutron neutral beam. Neutrons respond to boron inside the tumor and produce α-particles that destroy only tumor cells leaving intact the healthy cells. For this process to be effective, high concentration of 10B inside the cancer cells is needed. This can be achieved by the use of boron incorporated dendrimers and only if they can deliver it to cancer cells.

4.2   Therapeutics and Nanotechnology

Pharmaceutical sciences have a great interest in developing new nanotechnology products, understanding the new material behavior through the development of analytical assays and the possibility of measuring and taking pictures of products and devices in nanoscale. In the field of health, the problems of medicine's safety and efficacy, on time diagnosis and therapeutic confrontation of many diseases, like cancer, lead to the need of maximizing knowledge, aiming to the maximum development in therapeutic potential of the existing bioactive molecules [54].

The National Cancer Institute (NCI) and the US National Institutes of Health (NIH) under the Cancer Nanotechnology Project – a strategic initiative that targets the change in clinical oncology and basic research with immediate application of nanotechnology achievements – have completed the program by entering nanotechnology in biomedical research with encouraging results and started a new program applying nanotechnological achievements in cancer diagnosis, treatment, and prevention, with an ultimate goal of the reduction of deaths caused by cancer. The methodology for decision implementation includes finance securing for exchanging scientific opinions and researchers of all grades between universities and institutes in the fields of clinical oncology, molecular biology, and nanotechnology. The development of nanotechnology, electronics, and robotics is expected to offer important advantages in biomedical applications, like genome therapy, drug administration, imaging, and new drug discovery techniques.

Pharmaceutical nanotechnology orients in developing and evaluating bioactive molecule colloidal system transport, diagnostics, and imaging materials offering solutions in fields like:

- Imaging and diagnosis of diseases like cancer, in early stages.
- Developing systems for evaluating treatment effectiveness in real time.
- Multifunctional systems deliver bioactive molecules to damaged tissues in high concentrations.
- Nanomaterials that can detect molecular changes to prevent cancer metastasis or precancer cell transport. Innovative bioactive molecule delivery that can be used for central nervous system diseases with targeted imaging biomaterials.
- Uncontrolled cell multiplication and gene indicator (promoting cancer cells) detection systems.
- Research tool development that offer possibilities like new target cell identification and possible drug resistance detection.
- Study and development of new nanodimensional herbal medicinal products by improving their solubility in biological fluids.
- Alternative nanotechnological ways for insulin delivery.
- Applications in dental treatment by developing and evaluating new nanodimensional materials of high durability and biocompatibility.

For the pharmaceutical industry, these new technologies that deliver bioactive molecules are a strategic tool for their development. Technology can offer new solutions for the already existing bioactive molecules, increasing their effectiveness, improving their safety, and offering better patient compliance. Moreover, new drugs being developed under computational chemistry, using the knowledge gained from the human genome decoding program, need administration and delivery to the damaged tissues systems in order to be effective. Nanotechnology offers the possibility of administering bioactive molecules that are practically insoluble to water or unstable to biological environment. Designing and developing delivery nanosystems that target damaged tissues are an emerging direction in therapy. Even though nanotechnology offers the possibility of developing effective delivery systems for all kinds of bioactive molecules, the major directions are the developments in drug administration in diseases of the respiratory, the central nervous, and the cardiovascular system.

The advantages of nanotechnology in pharmaceutical sciences come as a result of controlling the nanomaterials' physicochemical properties and biological action. The cooperation of nanotechnology and biotechnology can lead to the development of nanodevices and nanosystems for studying the physiological processes of the living matter. The most important biomaterial in living organisms is their genetic material and mainly the DNA – also proteins and other biomolecules/biomaterials that participate actively in the existence of the living matter and its functions. The transport of the genetic material that is complexed with nanoparticles can offer solutions in genetic diseases and create appropriate conditions for the development of nanosystems that transfer and deliver genetic material with advantages in the genome therapy.

Nanoparticles consisted from polymers, synthetic or natural, from naturally occurring or synthetic lipids or from inorganic materials. Bioactive molecules that transfer nanoparticles can interact with them (covalent bonds or electrostatic forces) or encapsulate them in order to transfer them to the target tissues. The biomaterials mostly used for biodegradable nanoparticles and nanosystem production are polyacrylates, especially the quickly biodegradable butyl cyanoacrylate, gelatin, polylactic acid, and the polylactic acid copolymers.

Also, molecules of lipid nature that are compatible with human organism and biodegradable are used for nanosystem development, i.e., cholesterol, lipid acids, and phospholipids. The problem arising during bioactive molecule transport is their recognition from the MPS that detects them as foreign biomaterials and destroys them through opsonins' detection mechanism.

Opsonins are proteins functioning as flags, detecting the foreign-unknown biomaterials that are then detected from the macrophages of the MPS and are destroyed. Therefore, their long period of time stay in blood circulation that aims to the bioactive molecule transport to target cells is related to their surface physicochemical properties and their size. Changing the surface characteristics leads to behavior change and possibly reticuloendothelial system (RES) detection while particles of greater size like micro-sized systems or nanosized systems are detected and destroyed. Modern bioactive molecule nanocarriers have the ability of increasing the therapeutic index of the existing drugs by lowering the toxicity in tissues and achieving control release therapeutic levels for a prolonged period of time. Nanocarriers can increase the solubility of lipophilic bioactive molecules and improve their stability and therefore allowing the use of bioactive molecules that were rejected during pre-clinical and clinical studies due to short pharmacokinetics and biochemical and physicochemical properties. Lastly, nanotechnology can help in the development of multifunctional nanosystems that transfer bioactive molecules to the targeted area and nanosystems having diagnostic and therapeutic properties at the same time, allowing the simultaneous diagnosis and treatment of the damaged tissue. The multifunctional nanoparticle whose development in the future will offer the possibility of many independent functions. Nanosystems that can be characterized as nanorobotics are very difficult to be developed due to their complexity of their physicochemical properties and their *thermodynamic abnormalities* that lead to their instability. Despite all that, modern technology moves toward that direction. Target detection can occur in organ level or even in a single surface macromolecule target that is specific for some cells like surface antigens. The most common form of identification is the identification in molecular level that is based on the fact that each organ's or tissues' special macromolecules (antigens) can be detected. To achieve that targeting macromolecules can be used, i.e., antibodies capable to interact with a target expressing a special macromolecule (antigen). Nowadays, many protocols for the development and evaluation of innovative medicines are being developed and have various approaches. In some occasions, for damaged tissue targeting, many physiological characteristics of the targeted area can be used. In basic protocols of targeted therapy that have already been used experimentally or in clinical trials, the following approaches are included: the approach of the bioactive molecule to the damaged tissue; the passive accumulation through structural and functional abnormalities of the vessels that appear in areas with tumors, heart attacks, or inflammations; the natural target that is based on the difference of a physical parameter (i.e., pH or temperature) in the target area (such as tumors or inflammation); the magnetic targeting that is connected to paramagnetic carriers, with the effect of outside magnetic fields; and the active target using molecule carrier that are specially related to the damaged tissue or the cancer cell. There are already synthetic nanoparticles (liposomes, nanocapsules, nanospheres, dendrimers, etc.) that can target and enter damaged tissue cells releasing the bioactive molecules. Nowadays nanopharmaceutics and nanomedicine lead to the development of products that offer great advantages and effectiveness. The majority of the candidate bioactive molecules present low solubility in water, preventing their effectiveness. Research has shown that when the bioactive molecules are encapsulated in nanoparticles that have the appropriate physicochemical properties especially on their surface, preventing their aggregation, a stable and effective nanosystem carrier of the bioactive molecule is developed. The nanocrystal technology that big pharmaceutical industries in the United States apply has led to the development of therapeutic products [rapamycin (Rapamune® Pfizer) and aprepitant (Emend® Merck & Co.)] that are based on this technology.

The development of nanotechnology, electronics, and robotics is expected to offer great advantages in biomedical applications like gene therapy, bioactive molecule transport, imaging, and diagnostic techniques. Many bioactive molecules are ineffective due to limited access to target tissue. Nowadays, the major directions of pharmaceutical and medical research are therapeutics, lab diagnostics, and in vivo imaging diagnostics. There is an important effort in developing innovative medicines according to the interdisciplinary path of nanotechnology for the therapeutic approach for many diseases.

4.2.1 Nanotechnology and Cancer

During the past decades, there is an important progress in understanding and in description of the carcinogenesis mechanisms, while important diagnostic tools have been developed for damage tissue imaging and treatment. Despite this progress, the universal cancer mortality is one of the most important causes of death.

The in-depth knowledge of genetic and industrial changes that are responsible for cancer cell development has changed the therapeutic confrontation strategy. During the past years, new methods in diagnostics are developed that aim to the early malignant diagnosis and its therapeutic confrontation. Scientists have also understood that the microenvironment of cancer cells affects the treatment and provokes drug effectiveness.

Therefore, even if normal cells that neighbor cancer cells do not present changes in their genetic material; can change their physiological activity because they are surrounded and close to the cancer cells. Understanding the microenvironment around cancer cells, and not only their evolution and their behavior, is crucial in order to understand their development and design a strategy that scientists need to follow to develop effective anticancer agents targeting the cancer tumor and the environment where it grows. Moreover, a multifunctional system is developed that includes physical chemistry, biochemistry, and biophysics whose parameters should be evaluated in order to choose the therapeutic strategy. The factors and the procedures of this multifunctional system can be classified as follows:

- Factors related with the development and the multiplication of cancer cells and should be controlled.
- Physiological factors controlling the development and the multiplication according to the genetic information that cancer cells do not respond.
- Cancer cells do not obey the apoptosis procedure (programmed cell death).
- Development of vessel network process (angiogenesis) around tumor to provide oxygen and nutrients.
- Cancer cell metastasis process from the original tumor that results to death of 90 % of the patients.

It is scientifically proven that cancer therapy should be according to its complexity. The recognition of this complexity and the understanding of parameters and processes related to it have created new trends to the development of innovative medicines that are now designed according to the "system" cancer with the parameters and processes that define it. Nowadays, cancer therapeutics is oriented in the interdisciplinary cooperation and new technologies, with which we can detect sooner the "trace," understand the microenvironment, and design the bioactive molecule transfer, delivery, and target system. Dr. Andrew C. von Eschenbach, director of the National Cancer Institute (NCI), has set a target: to reach cancer therapy until 2015 or, more rationally, to develop effective therapeutic protocols. For this reason, the scientific alliance *NCI Alliance for Nanotechnology in Cancer* was founded in the United States that methodizes the tools and practices related to cancer prevention, diagnosis, and therapeutics with the means that nanotechnology can offer. This aims to bring in collaboration the physical, chemical, biological, and medicinal scientific community in a coordinated effort, so all nanotechnology benefits are directed to cancer patients [54].

NCI in the *NCI Alliance for Nanotechnology in Cancer* and other European-related initiatives emphasize the following sectors:

- Cancer prevention and control. Includes the nanodevice development for delivery of bioactive molecules to target tissues by using liposomes, dendrimers, nanoemulsions, etc. Complex anticancer vaccines will be synthesized using nanodevices for their systemic administration.
- Early diagnostics and proteomics. Implanted stable molecular sensors will be developed in order to find bio-indicators. These biosensors will be evaluated in situ or ex vivo, and the results will be transmitted through wireless technology to medicinal personnel and other databases.
- Imaging diagnostics. Aiming toward sensitive and precise imaging and "smart" injectable nanoparticles that will allow cancer tissue analysis cell by cell to be developed. Nanodevices analyzing the biological diversity of tumor's cancer cell will be also developed.
- Multifunctional therapies. There is a great need of nanosystems that will incorporate a combination of diagnostic and therapeutic functions. For example, nanocrystals can be used for both bioactive molecule transport and performing tissue imaging at the same time. Toward this direction, scientists are designing the development of smart devices in the near future that will control the bioactive molecule release in the particular place and time and evaluate the effectiveness of the treatment at the same time.
- Quality of life improvement during chemotherapy. Nanosystems for the treatment of nausea, pain, appetite loss, and fatigue will be designed.
- Interdisciplinary education. Nanotechnology success is relied upon the scientists' ability from various fields to interact and communicate effectively with each other. Steps like education for chemical engineers, physicists, and chemists on molecular and system biology as well as education of scientists on nanotechnology are very important.

The combination of the above scientific fields has led to the introduction of the term "nano-oncology" in the literature world [28]. Nano-oncology is divided into five basic fields: bio-imaging, gene therapy, thermal ablation, immunotherapy, and drug delivery.

The developments in managing genetic material and the scientific directions concerning its transport are related to the gene expression control during transcription, translation, and replacement of defective genes with compensatory genes. Despite the many possibilities of gene therapy, its clinical practice has important problems. More than 400 clinical studies have been evaluated during the past 15 years, and most of them have failed to achieve the desired results.

Here we can refer the liposomal vectors that are able to transport genes. It is well established that there are identified huge number of genes that are able to correct diseased phenotypes. Several delivery systems that are defined as nonviral vectors such as micelles, emulsions, and lipidic vectors have been used for gene delivery. A suitable nonviral vector should be stable and biocompatible to efficiently deliver genes to specific tissues, upon administration. The viral vectors despite their high efficiency in transfecting cells present a numerous problems like immunogenicity, toxicity, and difficulties in the scale-up process. However, the exploration of nonviral vectors is a demand. Liposomal vehicles could be systems that offer advantages because of their easy and safe administration by several routes such as i.p., i.v., etc. DNA can covalently attach on the surface of cationic liposome created complexes that are known as *lipoplexes* or can be accommodated into anionic liposomal vehicles. We have to keep in mind that the route of administration plays a key role in the in vivo results. Gene therapy is an emerging scientific field, and liposomes are considered as such nonviral vectors that could play a role in the development and evaluation process in gene therapy [45].

The effective gene therapy demands the combination of two different factors:

- A therapeutic gene that can be expressed in the target cell
- A safe – regarding the gene material safety – and effective delivery system that can transfer the gene in a specific tissue or organ

Even though impressive results have been published and presented when a therapeutic gene is injected on cancer tissue, its systemic administration has been proven to be especially difficult since most systems degrade in biological medium or excreted from the kidneys before reaching the target. Viruses, retroviruses, and adenoviruses that have presented infectious abilities are now abandoned because of the difficulty in incorporating into a large amount of genetic material and the safety matters related to carcinogenesis and immunostimulation. The synthesis of polymer nanosystems that transfer and deliver genetic material seems helpful in the development of innovative nanosystem that will be biocompatible, biodegradable, and effective. Some categories of polymer nanocarriers are the following:

- Poly-L-lysine.
- Synthetic biodegradable polycations.
- Chitosan.
- Cyclodextrins [55].
- Polymeric micelles (see Chap. 4). They have been used in research for gene transport and can be further divided in micelles composed of polyethylene glycol-polyester, polyethylene glycol-poly(amino acid), and polyethyleneglycol-polysaccharide lipid.
- Dendrimers.

Many studies have used dendrimers as gene material vectors through cell membrane to cell core. Liposomes and modified viruses have been extensively used for this reason. During the past years, PAMAM dendrimers have been tested and found to be very stable. They can carry great amount of genetic material in relation to viruses and are more effective than liposomes since they have a strictly defined structure and low pK amine values that help pH stabilization inside the human body. For this reason, PAMAM dendrimers tend to be established in an effective category of polycation synthetic dendrimers for gene transport.

To achieve local hyperthermia in order to destroy cancer cells is a subject that scientists study the past years but presents important problems. A basic problem occurring is the fact that the source of thermal energy might harm the surrounding healthy tissue even in case of targeted radiation. To resolve this problem, materials that selectively heat the tumor using gold nanoparticles absorbing near infrared have been developed. These systems are called nanoshells (see Chap. 3) and consisted of a silicon core surrounded by a thin gold shell. Nanoshells absorb heat (they are heated) during radiation in appropriate wavelength exposure. The characteristics of near infrared radiation have been chosen because the tissue absorbance in this wavelength is in minimum state while light permeation is in maximum state. To achieve hyperthermia, carbon nanotubes (see Chap. 3) have been used (tube structures with one carbon atom wall thickness) after the addition of specific antibodies. The results in breast cancer series were encouraging.

An important chapter in therapeutics is immune response. It has been proven that it can contribute in small tumor termination. But cancer cells have the ability to respond by developing mechanisms that provoke the organism's antigen ability. Research scientists try to develop immune system activation mechanisms against tumors that will possibly give the organism the ability to destroy cancer cells. Tumor antigens are not immune and if administered as vaccines can strengthen the immune system against cancer. The new vaccine toward this direction connects antigens with nanospheres. To increase the effectiveness, nanospheres must have a specific diameter (40–50 nm) and narrow size distribution that will allow them to target dendrimer cells of the lymph node.

The effectiveness of bioactive molecules that are administered to destroy cancer cells can be improved by using nanotechnological approaches. The design of appropriate nanosystems for bioactive molecule transport and delivery is an important research field that aims the development of *Trojan horses* that will not be detected from our immune system and will not affect healthy tissue cells. More in particularly, the following conditions must apply for the cytostatic bioactive molecules:

- Adequate bioactive molecule concentration in biological fluids will allow the effective concentration on the cancer tissue. For greater safety, the concentration of the bioactive molecule free form that is not loaded in the nanosystem must be the smallest possible in the biological fluids.
- The bioactive molecule should have high differential toxicity against cancer cells or at least a favorable therapeutic window.

Nanotechnology research aims to the points mentioned above using the special characteristics of cancer cells. These characteristics allow the passive and active nanosystem targeting cancer tumors.

High heterogeneity in cancer tumor vascularization includes areas of vascular necrosis and areas of high vascularization from where oxygen and nutrients reach cancer tissue. Cancer blood vessels present abnormalities regarding the corresponding normal blood vessels like high proportion of endothelial cells presenting abnormalities in the cell membrane, high blood vessel bending, and defects in the pericytes. Cancer capillary vessels present increased penetration that is mainly regulated by the abnormal secretion of the endothelial vessel growth factor, bradykinin, nitrogen monoxide, prostaglandins, and metal proteins.

Macromolecule transport in tumor microcirculation can take place through inter-endothelial connections and endothelial channels. The molecular exclusion level of these transport channels is less than 1 μm, and in vivo extravasation studies of liposomes in cancer xenografts have shown molecular exclusion of 400 nm. Generally particle extravasation is inversely proportional to size, and smaller particles (<200 nm) have shown to be more effective for tumor microcirculation extravasation. Lymph cancer network is also defective resulting in fluid retention and increased pressure relative to the fluid flow outside the tissue.

The absence of an intact lymphatic system results in nanocarrier's retention in the intracellular space. The combination of deficient microcirculation and the lack of an intact lymphatic system results into the enhanced permeation and retention (EPR) effect. Nanotechnology takes advantage of this effect for passive cancer targeting through nanosystem concentration into the cancer tissue at significantly higher levels compared with the plasma or healthy tissues. The bioactive molecule release from the nanosystem results in its relatively increased concentration in the tumor as well as the increased toxicity against cancer cells.

The modification of nanosystems' outside surface aiming the cancer tissue active targeting can be achieved by using antibodies, peptides, and small molecules that recognize special cancer antigens in tumor microenvironment. When these nanoparticles are directed at the outside section of the intramembrane cancer antigens, they are possibly taken over from the cancer cell through enhanced receptor endocytosis.

Known anticancer drugs like dimeric indole alkaloids, camptothecin, lignans derivatives, taxanes, and many more are the modern armory against cancer. Questions arising during treatment are the following:

- Why are current treatments not effective and most of the times fail?
- Why cannot we improve or completely eliminate drug toxicity and avoid toxic side effects?
- How can we achieve greater efficacy since the discoveries in the field of cancer molecular approach have noted gene expression in cancer cells leading to specialized protein production?

Current therapies fail to destroy compact tumors for three reasons:

1. At the state of diagnosis, the tumor is already well developed. A tumor of 1 cm³ size (the smallest clinically traceable tumor) contains one billion cells. To achieve

complete therapy, all cells must be destroyed. Even if we manage to destroy 99.9 % of those cells, one million of living cancer cells will remain in the organism.

2. Fifty percent of patients that will surgically remove the tumor will not be cured due to metastasis. During metastasis, the genetically modified cells will expand from their initial place and through circulation will be installed in new places like the liver and lungs. Usually, metastasis is not traceable due to their small size (<5 mm) and can remain inactive for many years.

3. The third and greatest obstacle in achieving successful treatment is tumor heterogeneity. Tumor contains cells with different genetic materials and biochemical, immune, and biological characteristics. Cells can vary according to surface cell receptors, enzymes, karyotype, morphology, recycle time, sensitivity in various therapeutic factors, and metastasis ability. This heterogeneity sets a limitation in surgery and therapy to achieve total tumor cell destruction. Apart from all the above, especially in the colon, kidneys, and adrenal gland tumors, the gene of P-glycoprotein is overexpressed resulting in tumor drug resistance.

Drug resistance is due to mrd-1 glycoprotein that carries cytotoxic drugs from the inside of the cell through adenosine triphosphate (ATP). It is believed that tumor cell compact structure is another obstacle for drug transportation into the tumor. Compact tumors do not have sufficient lymph system and as a result there is an elevated pressure in tumor's center. It is believed that the inside increased pressure in combination with the fast and abnormal tumor cell increase is responsible for the compaction and the blood vessel exclusion. After entering the organism, the bioactive molecule is not selectively accumulated in the damaged tissue but is distributed in various organs and tissues (depending on the bioactive molecule nature and administration route).

Despite all these, in order to reach its target (organ, tissue), it must come through many biological barriers, like other organs, cell membranes, and intercellular compartments where it can be disabled or present side effect in organs and tissues that are not involved in the pathological process. Therefore, in order to achieve the desired therapeutic concentration in the desired area in the organism, we must administer a great concentration of the bioactive molecule, which most of it will be lost in healthy tissues.

Furthermore, cytotoxic drugs have side effects as they cannot discriminate between healthy and tumorous cells (i.e., the most commonly used anticancer bioactive molecule, doxorubicin, presents cardiotoxicity). During the last decades, efforts are made to resolve the problems above through targeting of the bioactive molecules to the damaged tissues. Generally speaking, bioactive molecule targeting can be defined as the ability to accumulate into the organ target selectively and quantifically despite the route and method of administration. The bioactive molecule local concentration must be greater in the damaged area, while its concentration in the rest of the organs and tissues should not be above a certain level, preventing or limiting side effects.

For target therapy, the following conditions should apply:

• Drug administration protocols should be simple.
• The drug concentration needed for therapeutic result should be as small as possible, to prevent toxicity to healthy tissues.
• Bioactive molecule concentration in target tissue should have the ability to increase enough, without causing side effects in the rest of the tissues.

The theory of target therapy was introduced by Paul Ehrlich 100 years ago, and it involved a hypothetical *magic bullet* with two components: the first should detect the target and attach to it and the second should have the therapeutic effect.

Nowadays the theory of the *magic bullet* to target the damaged tissues involves different ingredients. The most common form of damaged tissue identification is the molecular level identification and is based on the fact that in every organ or tissue special macromolecules (antigens) can be found and only expressed in the particular organ. To achieve targeting, biomolecules/biomaterials can be used that are capable of selectively interacting with the target (e.g., special antibodies for the corresponding antigens).

Nowadays many protocols for targeted therapy are being developed and include a variation of approaches. The basic schemes of a therapeutic approach that have been already used in the lab or in clinical practice include the following:

- The bioactive molecule direct application at the area of the damaged organ (tissue)
- Its passive accumulation through the "incompatibilities-abnormalities" appearing in the vessels next to cancer tumors
- The target process that is based on a physical parameter (i.e., pH or temperature) at the target area (like tumor or inflammation)
- The magnetic targeting of nanosystems that are connected with paramagnetic materials, influenced by external magnetic fields
- The active targeting using molecule carriers that are specially related to the damaged tissue or cancer cell

There are many nanosystems (i.e., liposomes, nanocapsules, nanospheres) that can target cancer cells, enter inside their membrane, and release the bioactive molecules.

4.2.2 Nanobiotechnology

Nanobiotechnology is the interdisciplinary field of nanotechnology with biological systems and includes biophysics data that are described in Chap. 2. In the science of biology, many of the organism's biological structures and biomaterials have the same evolution and development with nanotechnology. The combination of these two fields is the interdisciplinary field of nanobiotechnology aiming at the nanosystem development from materials that possess both properties of nanomaterials and biomaterials, applied in disease therapeutics, the development of "green" energy, and the elimination of environmental pollutants.

The relation between biology and nanotechnology can be found in processes that are common in both scientific fields for the system development and the organization (bio- and nano-, respectively). The basic procedure reported in biological systems and nanosystems is self-assembly that controls various processes kinetically and thermodynamically. The most important biomaterial used by biomedical engineering, the scientific environment of nanobiotechnology in the field of health that combines the biological properties with the nanomaterial properties, is the genetic material, the DNA. DNA is implicated in numerous biological functions through biochemical paths while, because it is a nanomaterial, we can approach its physical properties through the science of biophysics. Recently, literature suggested that the genetic material plays an important biophysical role in physical phenomena with possible future biological applications. More in particularly, Kunming Xu [90] suggested at his work entitled *Stepwise Oscillatory Circuits of DNA Molecule* published in the *Journal of Biological Physics*, 2009 that the DNA biomolecule functions as a stepwise oscillatory circuit of electromagnetic radiation. One DNA molecule is characterized as a stepwise oscillatory circuit of electromagnetic radiation, where each base pair is a capacitor, each phospho group acts like an electric self-induction, and each desoxyribose acts as an electric switch. The circuit calculates the DNA conductance through charge rebounds between smaller and greater distances according to the experimental results that lead to many stages of mechanical rebounds. Therefore, in a charge rebound that is opposite to the phenomenon, the circuit acts according to the charge transfer mechanisms that reflect the genetic material credibility in electron transfer. The stepwise oscillatory charge transfer through DNA sequence is the one controlling the oscillatory frequency. Another approach of biophysical background concerns the electromagnetic signals that are developed from water nanostructures derived from bacterial DNA nucleotide sequences. Montagnier and his partners published a study presenting a new DNA property that is based in some bacterial sequence property to produce electromagnetic waves in high concentration water solutions. This appears to be a coordination phenomenon due to the low-frequency electromagnetic field wave that is applied. Most pathogenic organisms' DNA contains sequences that produce these signals. This phenomenon gives rise to the development of high-sensitivity detection systems for chronic bacterial infections in human and animal diseases. Most pathogenic organisms' DNA contains sequences that are able to produce these signals. This phenomenon raises new possibilities for the development of high-sensitivity tracking systems in both human and animal chronic bacterial infections. From all the examples mentioned above, the contribution of biophysics in the study of materials that structure the living organisms and are involved in numerous biochemical paths that lead to biological processes is observed. Classic physics, through its principles and laws, offers tools that can be used in understanding the behavior of these materials, not as chemical macromolecules, but as material sections with such a behavior that is based on their physical abilities like electrical conductance and electromagnetic radiation emission. Modern physics, especially quantum mechanics, has set a framework for the DNA behavior interpretation, one single nanobiomaterial that codes the genetic information.

4.2.3 Nanogenomics and Nanoproteomics

The term nanogenomics can be defined as the nanobiotechnology application in an organism's genetic material study. Some of the technologies used when studying the genetic material are presented in Table 4.4. Also, important technologies for the

Table 4.4 Examples of application of nanoparticles in gene therapy

Nanoparticle	Application
Poly(D,L-lactide-co-glycolide) nanoparticles with entrapped stem p53 DNA	Inhibition of cellular proliferation in cancers due to sustained expression gene with consequent release of intracellular p53
Intravenous administration of liposomal complexed form with composition of DOTAP: Chol-FUSI for repression of FUSI gene	Inhibition of tumor growth in mouse models with metastatic lung cancer
Cationic gelatin nanoparticles	Nonviral and nontoxic vectors for gene therapy
Calcium phosphate nanoparticles	Nonviral carriers for targeted therapy
Nonionic polymeric micelles composed of poly(ethylene oxide)-poly(propylene oxide)-poly(ethylene oxide)	Gene transfer (gene) in the gastrointestinal tract using oral administration in laboratory animals (mice)
Nanocomposite materials: titanium oxide nanoparticles in combination with DNA oligonucleotides which are activated by light or radiation	Genes encoding antibodies may be transferred to a particular intracellular target and in combination with radiation therapy aimed at killing cancer cells in patients
Combination of gene nanoparticles and surfactants	Gene transfer to the brain through the blood-brain barrier permeation
Integrin-targeted nanoparticles	Targeted delivery of anticancer drugs
DNA nanoparticles (20–25 nm): each DNA molecule covered with positively charged peptides	Crossing nanoparticles by nuclear passages (pores) with millions folds to facilitate gene expression compared to the non-genomic DNA. Used for trans-nasal treating cystic fibrosis
Nanoparticles' complexes with composition: EGF-PEG-biotin-streptavidin-PE-DNA	It presents great post-vaccine (transfection) effectiveness absence of aggregation of nanoparticles
Dendrimers with composition polyamino(amine) which can hold within DNA	Non-immunological carriers for in vivo gene delivery

study of the genetic material and its applications in therapeutics and diagnosis are mentioned in Table 4.4 and in the following section:

- Genetic polymorphism nanotechnological determination
- Nanoparticles for cancer therapy through p53 (immunolipoplex for p53 gene transfer)
- Silicon and gelatin nanoparticles for gene transport

With respect to nanogenomics, nanoproteomics is the application of nanotechnology in proteomics. The term proteomics expresses the protein effect study on absorption, distribution, metabolism, and excretion (ADME) from a bioactive molecule organism. The determination of protein effect with nanotechnology techniques is described as follows:

- Multiphoton detection of proteins
- Nanoflow liquid chromatography
- High-field asymmetric waveform ion mobility mass spectrometry

- Nanoproteomics for study of misfolded proteins
- Use of nanotube electronic biosensor in proteomics
- Nanofilter array chip detection

Some of the most common nanoparticles used to study the mitochondria are the following:

- Biosomes. Self-assembled structures composed of amphiphilic molecules (bola-amphiphile) that develop nanosystems for mitochondrial DNA transport in gene therapy.
- Liposomes are designed for bioactive molecule transport in the mitochondria using ligands with hydrophobic areas.
- Nanoparticles with special ligands that target mitochondria aiming to take control of mitochondrial functions.
- Quantum nanodots with special ligands that target the mitochondria aiming to take control of mitochondrial functions and morphology.

4.2.4 Nanotechnology and Biological Treatments

Biological therapies are those therapies where molecular biology can be applied. Biological therapies include vaccines, gene therapy, antisense, and RNA interference. Some of the biological therapies include the use of nucleic acids and proteins, while others include the genetic material management. The application of nanotechnology in therapeutics is definitive as materials and systems can be managed in nanoscale. Also, biological therapies use therapeutic products and not classical pharmaceutical products (bioactive molecules). The European Medicines Agency (EMA) (see Chap. 7) committee for therapeutic products' approval is called Committee for Advanced Therapies (CAT) (see Chap. 7).

The increased availability of therapeutic biological products in the market, like proteins, peptides, and antibodies, offers an important asset in therapeutics. These therapeutic macromolecules mentioned above have important advantages against conventional bioactive macromolecules, while the route of administration is an important research area for academic institutes and pharmaceutical industries.

Per os administration is an important drawback since these therapeutic products will permeate the surface of the buccal cavity or other biological membranes with difficulty. The sensitivity of their structure and conformation leads to peptide bond cleavage, proteolysis, oxidation, etc., and the disruption of noncovalent interactions resulting to aggregation, sedimentation, and finally immune response is developed. Problems mentioned above flag the macromolecule sensitivity in relation to the conventional bioactive molecules. They require special handling techniques, and the administration route should be chosen taking under consideration the rapid liver xxx that requires controlled doses.

It is obvious that their administration demands proper delivery systems that will enhance the above problems and will try to eliminate side effects. The alternative routes of administration are oral, nasal, and pulmonary. Today, the most common route of administration is the parenteral associated with compliance problems due to the repeated therapeutic dose, especially for chronic diseases. For example, diabetes is a disease that requires repeated doses and, therefore, long-term compliance that is a problem for patients. Biological therapeutic products transferred through mucosal and other administration routes, like oral, nasal, rectal, buccal, transdermal, and ocular, are under study. Tables 4.5 and 4.8 present biological therapeutic products that are in clinical trials or already in the market.

Systems mentioned in Tables 4.5, 4.6, 4.7, and 4.8 offer advantages in the field of therapeutics based on biological products. The development of micro- or nanosystems for drug delivery and administration through the routes mentioned above is

Table 4.5 Development of orally administered therapeutic biological products which are in clinical phases

Company	Product	Clinical phase
Emisphere Technologies	Salmon calcitonin	Clinical phase III
Emisphere Technologies/Novo Nordisk	GLP-1 (glucagon-like peptide)	Clinical phase I
Emisphere Technologies/Novo Nordisk	Insulin	Clinical phase I
Biocon	IN-105 (insulin conjugate)	Clinical phase III (India)
		Clinical phase I (USA)
Fosse Bio-Engineering Development Ltd.	Insulin	Clinical phase III
Generex	Oral-lyn™ (buccal insulin spray)	Approved product for sale on the market in many countries. Experimental products in the United States

Table 4.6 Developed inhaled technological forms of insulin

Company	Product	Clinical phase
Pfizer	EXUBERA®	Withdrawal due to low sales
Novo Nordisk	AERx®	Interruption in clinical phase III
Eli Lilly and Company and Alkermes Inc.	AIR® Insulin	Completion of the clinical phase III
MannKind Corporation	Technosphere® Insulin System	Clinical phase III (USA, Europe, Latin America)
Baxter	Recombinant human insulin inhalation powder (RHIIP) based on Baxter's proprietary PROMAXX formulation technology	Clinical phase I
Ventura and MicroDose Technologies Inc.	QDose insulin	Encouraging results of inhaled insulin will be announced. Prospective clinical studies

Table 4.7 Therapeutic biological products administered by the nasal cavity (nasal administration)

Company	Product	Description
Sanofi-Aventis	Kryptocur®	Nasal administration of luteinizing hormone-releasing Hormone (LHRH)
Novartis	Miacalsin®	Nasal administration of salmon calcitonin
Unigene Laboratories/Upsher-Smith Laboratories, Inc.	Fortical®	Nasal administration of salmon calcitonin
Ferring Pharmaceuticals, Inc.	Desmospray®	Nasal administration of desmopressin (analog of 8-arginine vasopressin (ADH)
Sanofi-Aventis	Suprecur	Buserelin (agonist of LHRH)
Sanofi-Aventis	Suprefact	Buserelin (agonist of LHRH)

a scientific and technological challenge oriented in their development from the pharmaceutical industry. Polymer chemistry and the design/evaluation of drug delivery systems (see Chap. 5) in nanoscale in order to enable biological product transfer will contribute in effectiveness and decreased side effects of these biological macromolecules.

4.2.5   Nanotechnology of Vaccines

The development of nanosystem transferring DNA for the vaccines' development and production, nanoemulsions (see Chap. 4), and nanoaerosols is an important direction in the field of vaccines. Table 4.9 presents examples of vaccines that are currently in the market or in different development stages and their administration is either nasal or per os. Vaccination is one of the most valuable and cost-effective health measures to prevent and control the spread of viral/bacterial infectious diseases responsible for high mortality and morbidity as reported in Strategic Research Agenda for Innovative Medicines Initiative 2 (IMI2) "The right prevention and treatment for the right patient at the right time" and in the report by Vaccine Europe 2013 entitled "Advancing health through vaccine innovation." A significant number of infectious diseases and chronic disorders are still not preventable by vaccination such as HIV, tuberculosis, malaria, healthcare-associated infections (HAIs), cytomegalovirus (CMV), and respiratory syncytial virus (RSV) for which new generation vaccines are needed. Novel technologies such as adjuvants (including immune modulators and molecular targets) can enable safe and effective vaccines for difficult target populations such as newborns, elderly, and the immunocompromised. The adjuvants need to be very well designed in order to avoid excessive response, long-term autoinflammatory diseases and allergy, and secondary effects. Therefore, particulates and especially liposomes could represent a perfect vaccine adjuvant, thanks to the possibility of a high level of customization and control. More recently,

Table 4.8 Development of micro- and nanosystems for mucosal administration of therapeutic biological products

Carrier	Size (nm)	ζ-potential (mV)	Loading method	Biomolecule	Loading (%)	Encapsulation efficiency
Oral delivery						
Chitosan	0.215	20.7	Encapsulation	Insulin	–	49.43
Chitosan			Encapsulation	DNA	–	
Chitosan/HPMCP	0.255	30.1	Encapsulation	Insulin	–	60.88
Chitosan/dextran sulfate	0.497–1.612	−21.5–3.2	Encapsulation	Insulin		48.6–96.4
Chitosan/dextran sulfate	0.527–1.577	−20.6–11.5	Encapsulation	Insulin	2.3	69.3
Chitosan/lecithin	0.121–0.347	7.5–32.7	Encapsulation	Melatonin	Up to 7.1	
DEAPA-PVA-g-PLLA	0.200–0.400	7.5–32.7	Self-assembly	Insulin	–	85
Polyacrylic acid/MgCl$_2$	0.278	−23.4	Encapsulation	Calcitonin	–	53.8
Lipid nanoparticles	0.200	−50.3	Encapsulation	Calcitonin	–	>90
Lipid nanoparticles/PEG	0.207–0.226	−36.6–34.8	Encapsulation	Calcitonin	–	>90
Nasal delivery						
Chitosan	0.275	46.7	Encapsulation	Insulin	44.1	46.9
Chitosan	0.040–0.600	18.8–31.1	Encapsulation	siRNA	13.2	9.2
Pulmonary delivery						
Liposomes	0.091–0.104		Encapsulation	Vasoactive intestinal peptide	0.4	
Liposomes/PEG	0.090–0.095		Encapsulation	Vasoactive intestinal peptide	0.4	
Lipid nanoparticles	0.115		Encapsulation	Insulin	–	

Table 4.9 Examples of vaccines on the market or in various development stages which are administered orally (oral administration) or by the nasal route (nasal)

Route of administration	Disease	Type of vaccine	Clinical phase
Oral	Poliomyelitis	Live attenuated	On market
	Typhus	Live attenuated	On market
	Cholera	Live attenuated	On market
	Acute gastroenteritis	Live attenuated	On market
	Diarrhea	Neutral/inactivated	Clinical phase III
	Dysentery	Live attenuated	Clinical phase III
	Ulcers in the digestive system, gastrointestinal cancer	Neutral/inactivated	Clinical phase I
	Anthrax	Vaccine strain	Preclinical phase
Nasal	Influenza	Live attenuated	On market
	Hepatitis B	Vaccine strain	Preclinical phase
	Diseases of the respiratory system	Live attenuated	Preclinical phase
	Anthrax	Neutral/inactivated	Preclinical phase
	Bronchiolitis/pneumonia	Neutral/inactivated	Preclinical phase
	Cervical cancer	Neutral/inactivated	Preclinical phase
	SARS	Vaccine strain	Preclinical phase
		Vaccine strain	

liposomes have found application as vaccine-adjuvants due to their ability to prevent antigen degradation and clearance, coupled with enhancing its uptake by professional antigen-presenting cells (APCs), and have marked liposomes as useful vehicles for the delivery of a diverse vaccine antigen. The majority of vaccines currently in development belong to the category of subunit vaccines, consisting of recombinant or purified pathogen-specific proteins or encoded (DNA) antigens that will be expressed and presented in vivo [66]. Subunit vaccines when administered alone have low efficacy in activating the immune system and require the addition of adjuvants in order to induce a measurable immune response of the antigen, through activation of the innate, and subsequently the adaptive immune system. Ideally, the adjuvant should be able to improve antigen uptake by APCs and induce an antigen-specific immune response while eliciting minimal toxicity [66]. Liposomes (Chap. 4) are a type of adjuvant that can potentially satisfy the above criteria. The adjuvant efficacy of cationic liposomes composed of Dimethyldioctadecylammonium bromide and trihalose dibehenate is well established in the literature. While the mechanism behind its immunostimulatory action is not fully understood, the ability of the formulation to promote a "depot effect" is under consideration. The depot effect has been suggested to be primarily due to the cationic nature which results in electrostatic adsorption of the antigen and aggregation of the vehicles at the site of injection. Virosomes are liposome-based vaccine formulations that are constructed from viruses without their genetic material. However, they are not able to replicate and to cause infection. Inflexal V (Berna Biotech Ltd.) and Epaxal (Janssen-Cilag Ltd.) are biological medicines that are vaccine-based antigen liposomal delivery nanosystems against influenza and hepatitis A virus (HAV), respectively. They are composed of DOPC (dioleoylphosphatidylcholine) and DOPE (dioleoylphosphatidylethanolamine), and their formulations are classified as suspensions. They mimic the viral infection promoting immune response while both are well tolerated and effective in children [14]. Dendrimers are under investigation as vaccine carriers and/or adjuvants for both infectious diseases and cancer immunotherapy. The adjuvant capacity of mannosylated poly(amidoamine) (PAMAM) dendrimers was documented in the literature as well as glycopeptides dendrimers and phosphorus dendrimers. Additionally, polymeric nanoparticles have been applied in vaccine delivery, showing significant adjuvant effects as they can easily be taken up by antigen-presenting cells.

4.2.5.1 Proteasomes™ as Vaccine Transport Vehicles

Proteasomes (Proteasomes™, GlaxoSmithKline, Brentford, Middlesex, UK) are considered to be vaccine transferring vehicles and create structures and cystic clusters at the size of viruses. The size of nanostructures is between 20 and 800 nm and depends on the type and the amount of antigen that is shaped into proteasome. The proteasome hydrophobic nature can contribute to vaccine transfer by facilitating the interactions between vaccine particles and their uptake from the ESN cells that will result in immune response. This technology is applied for vaccines against viruses, allergens, swines, and viruses of the respiratory syncytium.

Appendix

Liposome Preparation Protocols

Liposomes can be prepared with various techniques that offer energy in order for the phospholipid bilayers to bend and form pseudo-spherical structures.

The methods that are used for liposome preparation include the following basic steps: removal of the organic solvent that the phospholipids or lipids are dissolved, dispersed lipids in aqueous medium, encapsulation of the bioactive molecule into the lipid bilayers or into the aqueous core of liposomes, purification (e.g., using column chromatography) of liposomes encapsulated the bioactive molecule from its free from, and finally analytical methods to determine the encapsulation efficiency of the liposomal vehicles [60]. The conventional methods that are used for encapsulating the bioactive molecule into liposomes are:

- Reverse-phase evaporation technique [76]
- Ether injection technique [72]
- Freeze-thaw method [68]
- Rapid solvent exchange method [9]

There are two more conventional techniques that are used by the liposomal manufacturers:

- The French press technique [6]
- pH adjustment method [31]

The following liposome preparation protocols depend on the desired characteristics and size.

Thin-Film Hydration Method

This method (Fig. 4.15) is the most commonly used technique for multilamellar vesicles (MLVs). It is primarily based on lipid thin-film production that takes over the greatest possible surface in spherical flask in rotary evaporator under reduced pressure followed by hydration in a temperature greater than the main transition temperature of the phospholipids. The enclosed bioactive molecule to be incorporated or encapsulated is added either in the lipid film during its production if it is lipophilic or in the aqueous medium if hydrophilic [76].

The thin-film hydration method leads to multilamellar vesicles characterized of great size heterogeneity (1–5 μm). These liposomes can undergo size reduction and transform into small unilamellar vesicles (SUV) or large unilamellar vesicles (LUV) with greater size uniformity. This can be achieved by using probe sonication

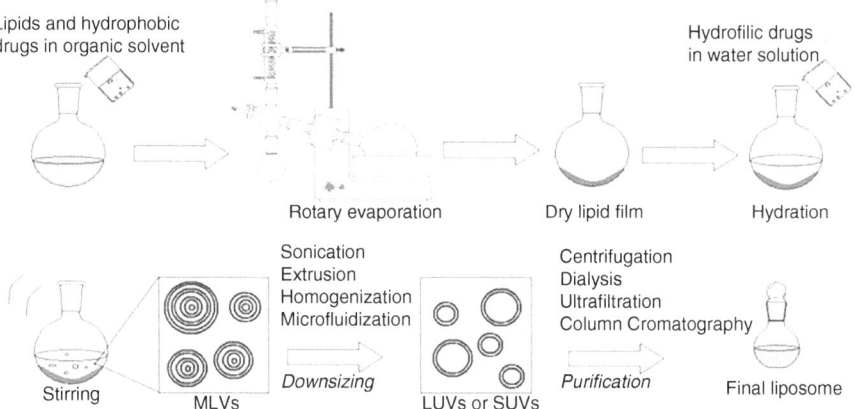

Fig. 4.15 Thin-film hydration method (Adapted from De Arou' jo Lopez et al. [2])

or extrusion through specific porous polycarbonate filters. Both sonication and extrusion provide energy to the system (thermal, mechanical) that is required for breaking the MLVs' liposome bilayers in small lipid vesicles, reducing the hydrophobic part exposure to the aqueous environment. Usually, before the use of polycarbonate filters, the liposomal MLVs' mixture undergoes fusion to reduce their size. The freeze-thaw method includes the rapid change between cold and hot temperatures of dry ice or liquid nitrogen and the basic phospholipid transition temperature used in liposome structural units. These extreme conditions break the bilayers up and produce smaller multilayer lipidic vesicles.

Solvent Injection

Phospholipid solvent injection in ethanol or diethyl ether in water medium is a technique for small or large unilamellar liposome (SUV, LUV) production. Phospholipids are primarily dissolved in a small volume of ethanol or diethyl ether and then quickly or slowly injected in water medium in a temperature greater than phospholipids' transition temperature.

The organic solvent is removed with evaporation, filtration, or dialysis. This method is simple and free from harmful (for the ingredients) chemical or physical processes, which is an important advantage of the method.

Nevertheless, the disadvantages of this method include the extra stage of the organic solvent and solute removal and the production of liposome dispersion systems.

Reverse-Phase Evaporation (REV)

Just like solvent injection method, phospholipids are dispersed in water environment through organic phase. The organic phase is not being removed but forms water in oil (w/o) emulsion. The phospholipid molecules are placed in unilamellars in a reverse micelle formation around the water molecules that during the organic solvent removal stage are agglomerated to form a mixture of unilamellar and multilamellar liposomes. Usually, a part of the water medium used for the liposome production is added into the solution of phospholipids in organic solvent. The rest is added after the organic phase removal. An advantage of this method is the great output in hydrophilic water molecules.

Heating Method

This method for liposome and nanoliposome production has been developed without using toxic chemicals and hazardous processes. This method includes liposome structural component hydration followed by their heating in a temperature of 120 °C in the presence of glycerol [56].

Glycerol is water soluble and physiologically acceptable from the human organism chemical molecule. It has the ability to increase lipid vesicle stability and doesn't need to be removed from the final liposomal product. The heating process is the major stage, and therefore, the method is called *heating method* and the produced liposomes are called heating method vesicles (HMV). By thermal method application, there is no need to sterilize the utensils used; therefore, the time for this method is reduced and the liposome production costs are limited. HMVs can be used as bioactive molecule delivery systems and for biological membrane simulation. In comparison to conventional liposome production, during this method, no volatile organic solutions are used, only glycerol that is a biocompatible and nontoxic parameter already used in pharmaceutical products preserving osmolarity during liposome production. There are cases where structures that are produced with *heating method* resemble the cell membranes of organisms in primordial earth. This is another leading indicator that liposomes are critical parameters in life's origin and determination.

Energy Demand and Approach of Liposome Preparation

To form lipid bilayers and, therefore, liposomal vesicles (i.e., liposomes), no matter the preparation method used, the hydrophilic/hydrophobic interactions between lipids and lipid/water molecules are essential. The energy offered (sonication or heat) in the lipid system results in lipid molecule appropriate orientation through hydrophobic interactions and, therefore, lipid vesicle formation to achieve the necessary thermodynamic balance of the lipid system. Lasic, in his book entitled *Liposomes: from Physics to applications* in 1993 [39], suggested that symmetric membranes favored flat configuration and energy should be offered for lipid bending.

The lipid type used and the presence/absence of cholesterol are parameters defining membrane rigidity. Whether *pseudo-spherical* lipid structures form liposomes in one or two stages (i.e., *pseudo-spherical* lipid vesicles), with specific thermodynamic content, as suggested by Lasic and coworkers, the basic condition for liposome formation is energy offered into the system. To summarize, liposomes are not formed spontaneously during lipid (mostly phospholipid) dispersion in aqueous medium, but extra energy input is required. In 'heating method,' for example, liposomes are formed due to energy offered into the system, and the mixing throughout the heating is performed to facilitate the homogenous ingredient distribution.

Polymersome Preparation Methods

During copolymer hydration, the motivation force for polymersome formation is the concentration gradient between copolymer diffused in water and water diffused into the copolymer. During simple hydration, the concentration degrades exponentially by time. The most common used method for polymersome formation is the thin-film hydration method. This method produces polymer emulsions with great size dispersion but also great amounts of other thermodynamically metastable polymersome structures.

Organic Solvent Method

This method is based on organic solvents and is related to water-in-oil-in-water (W/O/W) double emulsions. This method produces polymersomes with asymmetric membrane, where the inside and outside surfaces are of different nature. This is achieved by stabilizing water in oil (W/O) emulsion using the copolymer for the inside lamellae and producing vesicles permeating the water droplets through a second oil in water interface that consisted of the copolymer in order to form the outer membrane lamellae. Polymersomes are finally formed with controlled organic phase water removal using dialysis through membranes. The disadvantage of this method is that the organic solvent residues can cause biologic toxicity, limiting the produced polymersome application and use.

Dendrimer Synthesis

Many times during dendrimer production, scientists use, as cores bonds with special chemical and/or physicochemical properties, mostly photoactive or electroactive groups like porphyrins, ionic complexes, organ metal bonds, and fluorescent dies. The branches of dendrimers can be composed of repeated structural units. The number of the repeated structural units is controlled and defines the dendrimer generation (G). Each repeated structural unit is connected to the core or to another repeated structural unit through a branching spot. The branching spot between the core and the branch is of high importance since it determines the interaction between them. The branches play an important role in the dendrimer three-dimensional configurations and in the internal microenvironment. The dendrimer outer section ends up in peripheral groups. The part of these groups is also of great importance since they define the interaction between the dendrimer and each surroundings. The number of the peripheral groups increases exponentially as the dendrimer generation increases. Dendrimer formation always includes a series of repeated reaction sequence whose completion provides each time a new generation. Dendrimers are composed of monomers through polymerization and this process includes two methods. The methods can be classified into two basic synthesis methods, the divergent method (Fig. 4.16) and the convergent method (Fig. 4.17), and there are processes that combine these two methods (Fig. 4.18). The divergent method seems to be more effective for large-scale dendrimer synthesis, since the dendrimer weight doubles or triples in each generation. By increasing the generation number, the active group number in the surroundings increases exponentially by the necessary use of large excess of reagents. These reagents have a big difference in their molecular weight with the product and therefore are easily separated from it during reaction treatment. The exponential surrounding groups increase in each generation, increasing the possibility of incomplete reaction or by-product production. In these cases, the by-product separation is usually extremely difficult since they have similar weight

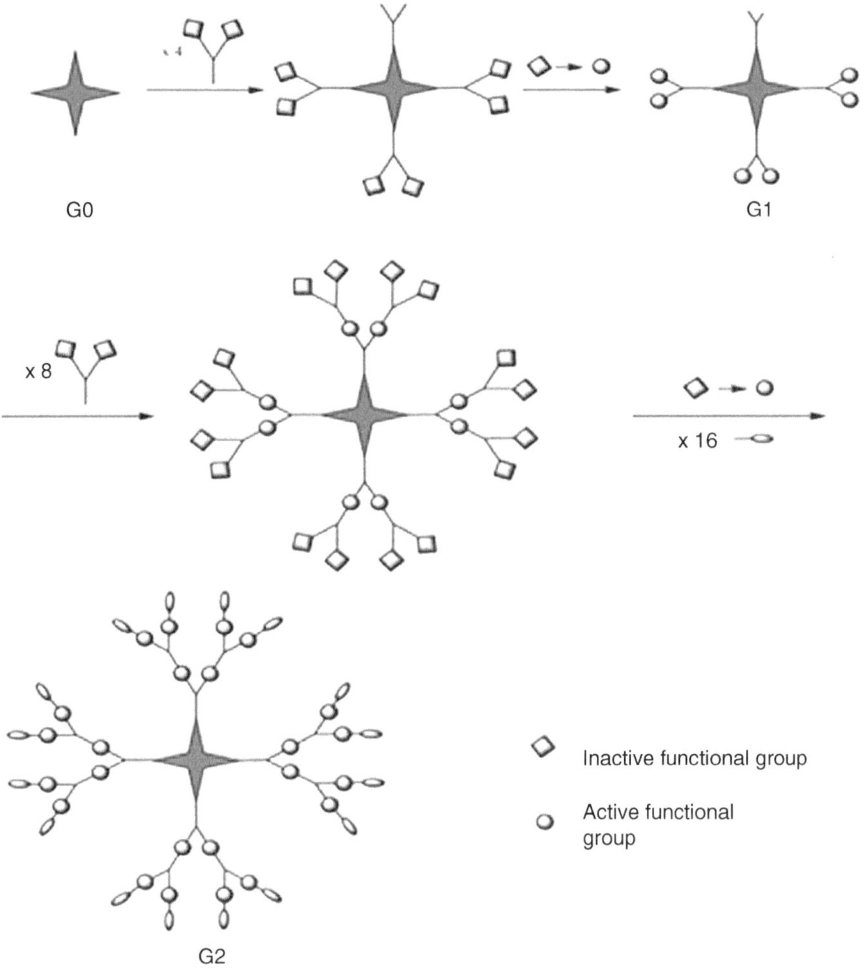

Fig. 4.16 Schematic of divergent synthesis of dendrimers (Adapted from [24] with permission from Bentham Science Publishers)

and structure with the dendrimer. The convergent reagent has more advantages in relation to the divergent method. Fewer reactions per molecule are needed during coupling and activation steps, and therefore, the use of large excess of reagents is not necessary – the cleaning is usually easier and the final product structure flaws are limited. Also, the precise surrounding group setting from the very first step allows the design of more target complex courses. In an effort of developing more efficient and less time-consuming complex courses for dendrimer synthesis, many complex courses combining the convergent and divergent method have been mentioned. This way the disadvantages of both methods will be limited by preserving their advantages. In 1991, Frechet's group [32, 89] mentioned a double-stage convergent

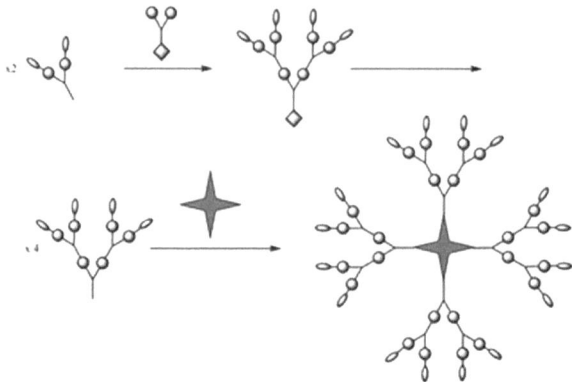

Fig. 4.17 Schematic of convergent synthesis of dendrimers (Adapted from [24] with permission from Bentham Science Publishers)

course (Fig. 4.18) in which the divergent method was used for large core formation (actually it was a low-generation dendrimer) followed by the convergent method for the coupling of the core and the branches that were formed separately.

Another combining approach aiming in complex course development with few steps is the orthogonal coupling method (Fig. 4.19) which targets in activation step removal during branching formation.

Summary

Liposomes are lipidic self-assembled nanostructures that are able to accommodate into their lipid bilayers or into their aqueous core, lipophilic and hydrophilic bioactive and biological molecules, respectively. Liposomes can be used as carriers for imagining and for diagnostic agents.

Liposomes are characterized as nanocolloidal lyotropic liquid crystals, and their *mesophases* that taking place during their thermotropic behavior affect their functions and their effectiveness as drug nanocarriers.

Liposomal membranes are studied based on their thermodynamics and biophysical aspects. Their thermotropic behavior affects their pharmaceutical effectiveness. They have been used to simulate biological membranes and to understand biological phenomena, i.e., the protein fusion in cell membranes.

Thermo- and pH-responsive liposomal vehicles are emerged nanotechnological platforms for drug delivery and targeting.

Polymersomes are composed of bilayers and their structural organization is close to that of liposomal bilayers.

Dendrimers belong to the very last generation of polymeric structures. They are considered as real nanoparticulate systems with very low polydispersity profile.

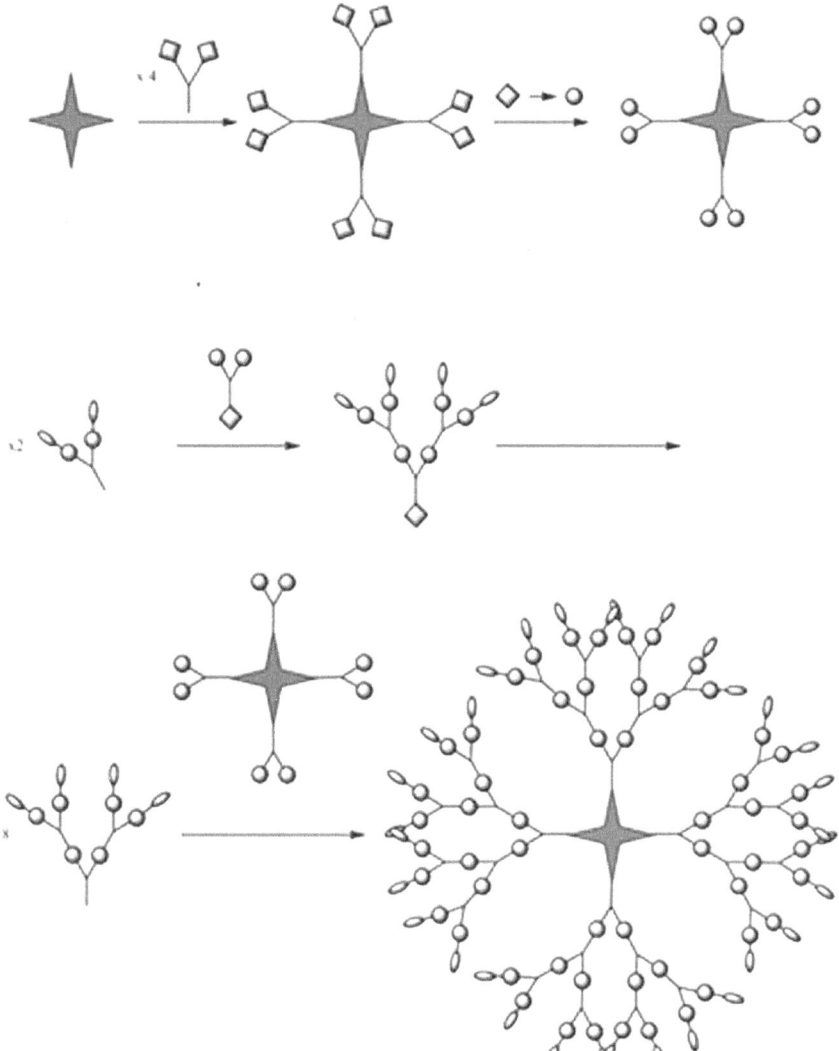

Fig. 4.18 Schematic of double-stage synthesis of dendrimers (Adapted from [24] with permission from Bentham Science Publishers)

They are used in diagnosis, in drug delivery, and as nonviral vectors in gene therapy.

Nano genomics is the term by which we describe as the nanobiotechnological applications.

Vaccines based on nanoparticulate systems are an emerging technology. Liposomes are promoted as a type of adjuvants that can potentially satisfy the criteria of adjuvanicity.

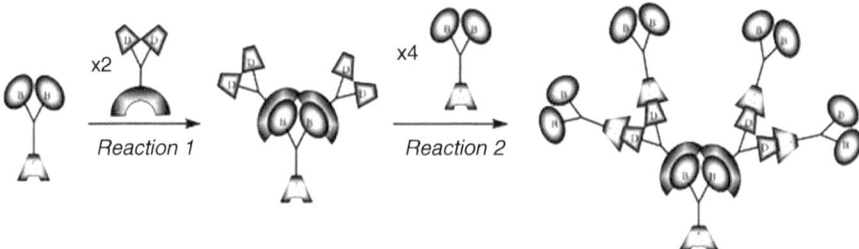

Fig. 4.19 Schematic of orthogonal conjunction synthesis of dendrimers (Adapted from [24] with permission from Bentham Science Publishers)

References

1. Allen T, Stuart DD (1999) Liposome pharmacokinetics classical, sterically stabilized, cationic liposomes and immunoliposomes. In: Janoff AS (ed) Liposomes rational design. Marcel Dekker, New York
2. De Arou' jo Lopez SC, dos Santos Giuberti C, Ribeiro Rocha TG et al (2013) Liposomes as carriers of anticancer drugs In: Rangel L (ed) Cancer treatment – conventional and innovative approaches. Rijeka, Croatia. InTECH ISBN 978-953-51-1098-9
3. Attwood D, Florence A (2012) Physical pharmacy. Pharmaceutical Press, Royal Pharmaceutical Society, UK
4. Aveling E, Zhou J, Lim YF, Mozafari MR (2006) Targeting lipidic-nanocarriers: current strategies and problems. Pharmakeftiki 19:101–109
5. Bangham AD, Standish MM, Watkins JC (1965) Diffusion of univalent ions across the lamellae of swollen phospholipids. J Mol Biol 13:238–252
6. Barenholz Y, Amselem S, Lichtenberg D (1979) A new method for preparation of phospholipid vesicles (liposomes)-French press. FEBS Lett 99(1):210–214
7. Becker AL, Henzler K, Welsch N et al (2012) Proteins and polyelectrolytes: a charged relationship. Curr Opin Colloid Interface Sci 17:90–96
8. Bertrand P, Jonas A (2000) Ultrathin polymer coatings by complexation of polyelectrolytes at interfaces: suitable materials, structure and properties. Macromol Rapid Commun 21:319–348
9. Buboltz JT, Feigenson GW (1999) A novel strategy for the preparation of liposomes: rapid solvent exchange. Biochim Biophys Acta 1417:232–245
10. Calvo P, Remunan-Lopez C (1997) Novel hydrophilic chitosan-polyethylene oxide nanoparticles as protein carriers. J Appl Polym Chem 63:125–132
11. Carrillo J-MY, Dobrynin AV (2011) Polyelectrolytes in salt solutions. Mol Dyn Simul Macromol 44(14):5798–5816
12. Cevc G (1996) Lipid suspensions on the skin. Permeation enhancement, vesicle penetration and transdermal drug delivery. Crit Rev Ther Drug Carrier Syst 13:257–388
13. Chenga R, Menga F, Denga C et al (2013) Dual and multi-stimuli responsive polymeric nanoparticles for programmed site-specific drug delivery. Biomaterials 34(14):3647–3657
14. Chang Hsin-I, Yeh Ming-Kung (2012) Clinical development of liposome-based drugs: formulation, characterization, and therapeutic efficacy. Int J Nanomedicine 7:49–60
15. Chu CJ, Szoka FC (1994) pH sensitive liposomes. J Liposome Res 4(1):361–395
16. Connor J, Yatvin MB, Huang L (1984) pH sensitive liposomes: acid-induced liposome fusion. Proc Natl Acad Sci U S A 81:1715–1718
17. D' Emanuele A, Jevprasesphant R, Penny J et al (2004) The use of a dendrimer-propanolol prodrug to bypass efflux transporters and enclose oral bioavailability. J Control Release 95:447–453

18. D' Emanuele A, Attwood D (2005) Dendrimer-drug interactions. Adv Drug Deliv Rev 57(15):2147–2162
19. D' Souza GGM, Boddapati SV, Weissig V (2006) Nanoparticulate carriers for drug and DNA delivery to mammalian mitochondria. Pharmakeftiki 19:110–121
20. Jhaveri A, Deshpande P, Torchilin V (2014) Stimuli-sensitive nanopreparations for combination cancer therapy. J Control Release 190:352–370
21. Fait JP, Mecozzi S (2008) Nanoemulsions for intravenous drug delivery. In: de Villiers MM, Aramwit P, Known GS (eds) Nanotechnology in drug delivery, vol X. Springer, AAPS Press, New York, pp 461–489
22. Felice B, Prabhakaran MP, Rodríguez AP, Ramakrishna S (2014) Drug delivery vehicles on a nano-engineering perspective. Mater Sci Eng C Mater Biol Appl 41:178–195
23. Frechet JMJ (1994) Functional polymers and dendrimers-reactivity, molecular architecture and interfacial energy. Science 263:1710–1715
24. Gardikis K, Micha-Screttas M, Demetzos C, Steele BR (2012) Dendrimers and the development of new complex nanomaterials for biomedical applications. Curr Med Chem 19:4913–4928
25. Gregoriadis G, Ryman BE (1972) Fate of protein-containing liposomes injected into rats an approach to the treatment of storage disease. Eur J Biochem 24:485–491
26. Gregoriadis G, Ryman BE (1972) Lysosomal localization of β-fructofuranosidase-containing liposomes injected into rats. Biochem J 129:123–133
27. Gregoriadis G, Buckland RA (1973) Enzyme-containing liposomes alleviate a model for storage disease. Nature (London) 2:170–172
28. Gregoriadis G, Willis EJ, Swan CP (1974) Drug-carrier potential of liposomes in cancer chemotherapy. Lancet 1:1313–1316
29. Gregoriadis G (ed) (1988) Liposomes as drug carriers. Wiley, London
30. Hartig SM, Greene RR, DasGupta J et al (2007) Multifunctional nanoparticulate polyelectrolyte complexes. Pharm Res 24:2353–2369
31. Hauser H, Gains N, Mueller M (1983) Vesiculation of unsonicated phospholipid dispersions containing phosphatidic acid by pH adjustment: physicochemical properties of the resulting unilamellar vesicles. Biochemistry 22:4775–4781
32. Hawker CJ, Lee R, Frechet JMJ (1991) One-step synthesis of hyperbranched dendritic polyesters. J Am Soc 113:4583–4588
33. Hong K, Kiprotin DB, Park GW et al (1999) Anti-her 2 immunoliposomes for targeted drug delivery. Ann N Y Acad Sci 886:293–296
34. Ihre HR, de Jesus PLO, Szoka FC (2002) Polyester dendritic systems for drug delivery applications: design, synthesis and characterization. Bioconjug Chem 13:443–452
35. Jenkins AD, Kratochvil P, Stepto RFT et al (1996) Glossary of basic terms in polymer science (IUPAC Recommendations 1996). Pure Appl Chem 68(12):2287–2311
36. Kitchens KM, Ghandehari (2009) PAMAM dendrimers as nanoscale oral drug delivery systems. In: de Villiers MM, Aramwit P, Kwon GS (eds) Nanotechnology in drug delivery, vol X. Springer, AAPS Press, New York, pp 421–460
37. Kono K, Murakami T, Yoshida T, Haba Y et al (2005) Temperature sensitization of liposomes by use of thermosensitive block copolymers synthesized by living cationic polymerization: effect of copolymer chain length. Bioconjug Chem 16:1367–1374
38. Kyrikou I, Daliani I, Mavromoustakos T et al (2004) The modulation of thermal and dynamic properties of vinblastine by cholesterol in membrane bilayer. Biochim Biophys Acta Biomembr 1661(1):1–8
39. Lasic DD (1993) Liposomes: from physics to applications. Elsevier Publishing Company, Amsterdam
40. Lasic DD, Barenholz Y (eds) (1996) Non-medical applications of liposomes. Press CRC, Boca Raton
41. Lasic DD, Papahadjopoulos D (1995) Liposomes revisited. Science 267:1275–1276
42. Lasic DD, Papahadjopoulos D (eds) (1998) Medical applications of liposomes. Elsevier, Amsterdam

43. Lee Y, Kataoka K (2009) Biosignal-sensitive polyion complex micelles for the delivery of biopharmaceuticals. Soft Matter 5:3810–3817
44. Levacheva I, Samsonova O, Tazina E et al (2014) Optimized thermosensitive liposomes for selective doxorubicin delivery: formulation development, quality analysis and bioactivity proof. Colloids Surf B Biointerfaces 121:248–256
45. Li S, Huang L (1999) Functional pleomorphism of liposomal gene delivery vectors. Lipoplex and lipopolyplex. In: Janoff AS (ed) Liposomes rational design. Markel Dekker, New York
46. Liu X, Huang G (2013) Formation strategies, mechanism of intracellular delivery and potential clinical applications of pH-sensitive liposomes. Asian J Pharm Sci 8:319–328
47. Liu J, Huang Y, Kumar A et al (2014) pH-sensitive nano-systems for drug delivery in cancer therapy. Biotechnol Adv 32:693–710
48. Mabrook E, Cuvelier D, Brochard-Wyrat F et al (2012) Polymersomes in polymersomes: multiple loading and permeability control. Angew Chem Int Ed Engl 28:11215–11224
49. Marquez-Beltran C, Castaned L, Enciso-Aguilar et al (2013) Structure and mechanism formation of polyelectrolyte complex obtained from PSS/PAH system: effect of molar mixing ratio, base-acid conditions and ionic strength. Collied Polym Sci 291(3):683–690
50. Massignani M, Lomas H, Battaglia C (2010) Polymersomes: a synthetic biological approach to encapsulation and delivery. In: Caruso F (ed) Modern techniques for nano- and microreactors/-reactions advances in polymer sciences, vol 229. Springer, Berlin, p 115
51. Matsingou C, Demetzos C (2007) The perturbing effect of cholesterol on the interactions between labdanes and DPPC bilayers. Thermochem Acta 452(2):116–123
52. Matsingou C, Demetzos C (2007) Calorimetric study on the induction of interdigitated phase in hydrated DPPC lipid bilayers by bioactive labdanes and correlation to their liposomal stability. The role of chemical structure. CPL 145(1):45–62
53. Matsingou C, Demetzos C (2007) Effect of the nature of the 3b-substitution in manoyl oxides on the thermotropic behavior of DPPC lipid bilayers and on DPPC liposomes. J Liposome Res 17(2):89–105
54. Mauro F (2005) Cancer nanotechnology: opportunities and challenges. Nat Rev Cancer 5:161–171
55. Moya-Ortega MD, Alvarez- Lorenzo C, Concheiro A (2012) Cyclodextrin-based nanogels for pharmaceutical and biomedical applications. Int J Pharm 428(1):152–163
56. Mozafari MR (2005) Liposomes: an overview of manufacturing techniques. Cell Mol Biol Lett 10(4):711–719
57. Mourelatou E, Spyratou E, Georgopoulos A et al (2010) Development and characterization of oligonucleotide-tagged dye-encapsulating EPC/DPPG liposomes. J Nanosci Nanotechnol 10:1–9
58. Muller RH, Mader K, Gohla S (2000) Solid Lipid nanoparticles (SLN) for controlled drug delivery – review of the state of the art. Eur J Pharm Biopharm 50(1):161–177
59. Nallani M, Andreasson-Ochsner M, Tan CV et al (2011) Proteopolymersomes: in vitro production of a membrane protein in polymersomes membrane. Biointerfaces 6(4):153–157
60. New RCC (1990) Liposomes a practical approach. IRL, Oxford University Press, Oxford
61. Newkome GR, Yao Z, Baker GR (1985) Cascade molecules: a new approach to micelles. A[27]-arborol. J Org Chem 50(11):2003–2004
62. Omri A, Agnew BJ, Patel GB (2000) Short term repeated dose toxicity profile of archaeosomes administered to mice via intravenous and oral routes. J Liposome Res 10:523–538
63. Papahadjopoulos D, Witkins JC (1967) Phospholipid model membranes II. Permeability properties of hydrated liquid crystals. Biochim Biophys Acta 135:639–652
64. Papahadjopoulos D, Vail WJ, Jacobson K et al (1975) Cochleate lipid cylinders: formation by fusion of unilamellar lipid vesicles. Biochim Biophys Acta 394(3):483–491
65. Papahadjopoulos D (ed) (1978) Liposomes and their use in biology and medicine. Ann N Y Acad Sci 408:1–412
66. Perrie Y, Obrenouin M, McCarthy D et al (2002) Liposomes (*lipodine*) – mediated DNA vaccination by the oral route. J Liposome Res 12(1–2):185–197

67. Photos PJ, Bacakova L, Discher B et al (2003) Polymer vesicles in vivo: correlations with PEG molecular weight. J Control Release 90:323–334
68. Pick U (1981) Liposomes with a large trapping capacity prepared by freezing and thawing of sonicated phospholipid mixtures. Arch Biochem Biophys 212:186–194
69. Pippa N, Karayianni M, Pispas S, Demetzos C (2015) Complexation of cationic-neutral block polyelectrolyte with insulin and in vitro release studies. Int J Pharm 491(1–2):126–143
70. Pispas S (2011) Self-assembled nanostructures in mixed anionic-neutral double hydrophilic block co-polymer/cationic vesicle-forming surfactant solutions. Soft Matter 7:474–482
71. Ruysschaert AFP, Sonne T, Haefele T et al (2005) Hybrid nanocapsules: interactions of ABA block copolymers with liposomes. J Am Chem Soc 127:6242–6247
72. Schieren H, Rudolph S, Finkelstein M et al (1978) Comparison of large unilamellar vesicles prepared by petroleum ether vaporization method with multilamellar vesicles: ESR, diffusion and entrapment analyses. Biochim Biophys Acta 542:137–153
73. Sidone B, Zamboni A, Zamboni W (2011) Meta analysis of the pharmacokinetic variability of liposomal anticancer agents compared with non liposomal anticancer agents. In: Abstracts of 2011 ASCO Annual Meeting. J Clin Oncol 29:2011 (suppl, abstract 2583)
74. Simões S, Moreira J-N, Fonseca C et al (2004) On the formulation of pH-sensitive liposomes with long circulation times. Adv Drug Deliv Rev 56:947–965
75. Slingerland M, Guckelaar HJ, Gelderblom H (2012) Liposomal drug formulations in cancer therapy: 15 years along the road. Drug Discov Today 17:160–166
76. Szoka FJ, Papahadjopoulos D (1980) Comparative properties and methods of preparation of lipid vesicles (liposomes). Annu Rev Biophys Bioeng 9:467–508
77. Ta T, Porter TM (2013) Thermosensitive liposomes for localized delivery and triggered release of chemotherapy. J Control Release 169(1–2):112–125
78. Taubert A, Napoli A, Meier W (2004) Self-assembly of reactive amphiphilic block copolymers as mimetics for biological membranes. Curr Opin Chem Biol 8(6):598–603
79. Tibaldi JM (2012) Evolution of insulin development: focus on key parameters. Adv Ther 29:590–619
80. Tomalia DA, Baker H, Dewald J et al (1985) A new class of polymers: starburst-dendritic macromolecules. Polym J 17:117–132
81. Tomalia DA (1994) Starburst/cascade dendrimers: fundamental building blocks for a mew macroscopic chemistry set. Adv Mater 6:529–539
82. Tomalia DA (2006) Dendrons/dendrimers quantized nano-element like building blocks for soft-soft and soft-hard nano-compound synthesis. Soft Matter 2:478–498
83. Tomalia DA (2009) In quest of a systematic framework for unifying and defining nanoscience. J Nanopart Res 11:1251–1310
84. Tompson KL, Chambon P, Versber R (2012) Can polymersomes form colloidosomes? J Am Chem Soc 134(30):12450–12453
85. Torchillin V, Trubetskoy VS (1995) In vivo visualizing of organs and tissues with liposomes. J Liposome Res 5:795–812
86. Torchilin VP (2004) Targeted polymeric micelles for delivery of poorly soluble drugs. Cell Mol Life Sci 61(19–20):2549–2559
87. Villalonga-Barber C, Micha-Skretta M, Steele BR, Geaorgopoulos A, Demetzos C (2008) Dendrimers as biopharmaceuticals: synthesis and properties. Topics Med Chem 8(4):1294–1309
88. Weiner N, Lieb L (1998) Developing uses of topical liposomes: delivery of biologically active macromolecules. In: Lasic D, Papahadjopoulos D (eds) Medical applications of liposomes. Elsevier, Amsterdam
89. Wooley KL, Hawker CJ, Frechet JMJ (1991) Hyperbranched macromolecules via a novel double-stage convergent growth approach. J Am Chem Soc 113:4252
90. Xu K (2009) Stepwise oscillatory circuits of DNA molecule. J Biol Phys 35:223–230
91. Zhang H, Gong W, Wang ZY et al (2014) Preparation, characterization, and pharmacodynamics of thermosensitive liposomes containing docetaxel. J Pharm Sci 103(7):2177–2183

Part III
Regulatory Framework

Chapter 5
Application of Nanotechnology in Modified Release Systems

Abstract The most appropriate route of administration is considered to be oral, because of the patient compliance and of economical issues. Drug delivery nanosystems are defined as technological platforms that promote the effective administration of bioactive molecules or therapeutic agents (protein, peptide, antibody, genetic material) in the human organism. Drug delivery nanosystems have been evaluated based on their physicochemical and structural properties and on the way that they behave within biological media. They are developed to optimize the production of new medicines, to ameliorate patient compliance, and to improve their targetability from systemic to specific tissues and cells. Hybrid and chimeric drug delivery nanosystems are two major classes that are categorized based on the nature of the mixing biomaterials to produce the final nanocarrier. Their ability to mimic the functions of natural objects proceeds to the category of bio-inspired drug delivery systems.

Keywords Drug delivery nanosystems • Routes of administration • Controlled release • Targeting • Bio-inspiration and bioengineering

5.1 Drug Delivery Nanosystems

The routes for drug administration could be categorized in two main sectors. The first one is the local and the other is the systemic. The systemic administration is further classified in the parenteral (intravenous, subcutaneous, and intramuscular) and enteral (buccal, oral, rectal, and sublingual) routes. The most appropriate route of administration among others is considered to be the oral, which is the most studied one, because of the patient compliance and of economical issues [16]. The oral route to administrate drugs is obviously the most adequate, but the oral delivery of biomolecules (i.e., proteins and peptides) for therapeutic purposes is hindered due to their low bioavailability. The enzymatic degradation into the gastrointestinal tract (GI) and the hydrolysis into the stomach hinder their administration orally, promoted the parenteral route. However, new and innovative drug delivery systems emerging as an approach could be contributed to the difficulties in drug delivery, especially of therapeutic biomolecules by developing nonviral colloidal systems with improved physicochemical properties that are able to transport biomolecules (i.e., proteins and peptides) to the desired tissues [36].

Primary research in drug delivery started in the 1950s by using polyclonal antitumor antibodies in order to specifically approach tumors with anticancer drugs [23]. There were published several efforts in the 1960s to deliver macromolecules using microcapsules [5]. It is important to figure out that the main concerns in drug delivery systems were on the one hand their physicochemical and structural properties based on the nature and behavior of their building biomaterials and on the other the way that they behave within biological media. Drug delivery systems are developed to optimize the production of new medicines, to ameliorate patient compliance, and to improve their targetability from systemic to specific tissues and cells [27]. New approaches in biology, immunology, biochemistry, and technology lead to produce innovative delivery systems that are able to control the pharmacokinetic and pharmacodynamic (PK/PD) behavior of bioactive molecules [34].

Nanotechnology offers benefits to the research, development, and scale-up processes for production of delivery nanosystems. The innovation is based on the multidisciplinary approaches in order to fabricate *smart* nanoscaled drug delivery systems. Drug delivery nanosystem is defined as a technological nanodevice that contributes in the bioactive molecule or therapeutic agent (protein, peptide, antibody, genetic material) effective administration in the human organism. As a result, the improvement of pharmacological effectiveness, the reduction of side effects, and the control of the pharmacokinetic parameters appear. The benefits from the bioactive molecule delivery nanosystem production are based not only on their properties because of their size but in their ability to achieve the following: tissue and organ targeting; bioactive molecule or therapeutic agent biochemical protection, with increased stability and lower toxicity within the organism; absence of unwanted interactions with the biological environment; great circulation time; substance biodegradability; easy and low cost production; and final product great storage time and physicochemical stability.

At this point, it is important to mention that, in nanoscale level, the biomaterials and final nanosystem results and properties are very important due to the surface over volume fraction that is greater in relation to microscopic system scale and to the quantum effects.

The pharmacokinetic analysis based on clearance offers quantitative information regarding biodistribution that is related to delivery nanosystem physicochemical abilities. Various drug delivery nanosystems – including liposomes, dendrimers, and polymersomes (see Chap. 4) – have been designed and developed until today [15]. The bioactive molecule delivery can be improved by using colloidal nanocarriers that include apart from the above, polymeric nanoparticles, micelles, and carbon nanotubes.

The bioactive molecules incorporated in nanosystem transport and delivery to damaged target tissues have two major advantages: the bioactive molecule is protected from the outer environment until it is released to the damaged tissue and the pharmacological effectiveness of the bioactive molecule is defined more from the nanoparticle nature and less from the bioactive molecule special physicochemical

Table 5.1 Basic physicochemical properties of drug delivery nanosystems

Regarding to their use	Regarding to their preparation and properties
Surface area	Interactions between drug and nanosystems
Shape	Interactions between different biomaterials of the nanosystem
Drug release	Hydrophobicity and hydrophilicity of surfaces of nanosystems surfaces
Diffusion (i.e., tissue and mucus)	Size, size distribution, and ζ-potential
Destabilization of nanosystems. Aggregation and/or flocculation (in vitro and in vivo)	Adsorption of macromolecules onto surfaces
Absorption (i.e., into biological and polymeric membranes)	
Deposition (i.e., into surfaces)	
Rheology (i.e., into blood)	
Flexibility (to help the movement into capillary networks)	

characteristics. This happens because the organism identifies, mostly through the immune system, the delivery nanosystem and not the bioactive molecule. This fact has extreme advantages in the effectiveness of therapy through nanotechnological approach for new medicine development.

The basic physicochemical properties of drug delivery nanosystems are presented in Table 5.1.

It is important to note that the difference between the pharmacokinetics of the bioactive molecule released from the nanosystem and the kinetics of the nanosystem within the organism is an important parameter affecting the final product effectiveness and should be studied. The rational design of a drug delivery nanosystem demands control and tests to preserve the stability characteristics after administration in the organism. The kinetics of nanosystems within the organism requires adjustments of their physicochemical characteristics – which should be anticipated during experimental design – through their limits to preserve their stability characteristics. Since nanosystems move through tissues, many changes take place in their environment, one of which is the change in the surface characteristics caused by opsonization process. These proteins that are attached to the "unknown" of the organism particle surfaces function as "flags" by detecting them and activating the macrophage cells of the endothelial system in order to destroy them.

5.1.1   *Activation-Modulated Controlled Release Delivery Systems*

In this controlled release system, the bioactive molecule release is activated from physical, chemical, and biological processes or is backed up from external energy. The enclosed bioactive molecule release rate is indirectly controlled by controlling

Table 5.2 Methods applied for the release activation of the loaded bioactive molecule of the activation controlled drug delivery systems

Systems activated by		
Physical methods	Chemical methods	Biochemical methods
Osmotic pressure	Change of pH	Enzymes
Hydrodynamic pressure	Change of ion concentration	Biochemical reactions
Vapor pressure	Hydrolysis	–
Mechanic energy	–	–
Magnetic energy	–	–
Ultrasound	–	–
Iontophoresis	–	–
Hydration	–	

the activation process or the energy delivered to the system. Table 5.2 presents the methods applied for the release activation of the loaded bioactive molecule of the activation controlled drug delivery systems.

5.1.1.1 Feedback-Regulated Controlled Release Systems

These are controlled release systems third generation, where the molecule release is regulated from a biochemical-induced biosensor. This sensor usually detects a biochemical substance in the organism that activates the system. Depending on this substance/activator concentration, the system feedback mechanism regulates the loaded bioactive molecule release rate. This specific controlled release technology is currently being developed and expected to be used in applications demanding precise and repeated bioactive large molecules doses, like insulin delivery in diabetic patients. Important feedback-regulated controlled drug delivery systems are the following:

- Bio-responsive systems
- Self-regulated systems

5.1.1.2 Triggered Controlled Release Systems

In these systems, the bioactive molecule is carried in a device surrounded from a biodegradable polymeric membrane. The membrane permeability is controlled from a biochemical substance (e.g., glucose) concentration in the tissue where the system is placed. A typical example of this diffusion-controlled or biodegradation-controlled release system is the glucose-triggered insulin delivery system. The insulin tank is in a hydrogel membrane with $-NR_2$ groups. In alkaline solution, $-NR_2$ groups are neutral and the membrane remains not swellable and not insulin permeable. Since glucose, the system biochemical activator, permeates the membrane, it is oxidized from glucose oxidase enzyme that is located in the membrane and forms gluconic acid. Then, the -NR_2 groups accept protons to form -NR_2H^+ and the hydro-gel membrane swells and renders permeable to insulin molecules. The amount of insulin released in the organism is proportional to the glucose concentration pene-trating the system membrane.

5.1.1.3   Self-Regulated Controlled Release Systems

These systems are based on a competitive binding system that activates and regu-lates the bioactive molecule release in the organism. The bioactive molecule is stored in a grid surrounded by a semipermeable membrane, and its release is acti-vated when a biochemical substance from the nearby tissue penetrates the mem-brane. One of the first examples of self-regulated delivery systems used a competitive binding system between sucrose molecules and lectin. An insulin-sucrose-lectin complex is placed in a semipermeable membrane. While blood glucose is diffused in the apparatus, it connects competitively on sucrose connective spots on lectin molecules (concanavalin A). At the same time, the insulin-sucrose molecules are released. Then the released complexes are diffused out of the appliance toward the blood.

5.2   Innovative Drug Delivery Nanosystems

Controlled release of a bioactive molecule from an innovative drug delivery system results in modulating the pharmacokinetic parameters and aims toward treatment effectiveness. The novelty is not only in the delivery nanosystem in relation to bio-materials composition, in the structural biomaterial cooperativity and in its thermo-dynamic characteristics, but in the possibility of modulating the loaded bioactive molecule release rate. The controlled release, basically the release rate, is related to concepts like bioavailability, pharmacokinetics, toxicity, and final therapeutic result of the final medicinal product. Liposomes, polymeric nanosystems, dendrimers, etc. mentioned in Chap. 4 are classified as controlled release therapeutic nanosystems.

   It is estimated that the majority of the bioactive molecule candidates that are studied to become drugs are either hydrophilic or lipophilic regarding their physico-chemical characteristics. Low water solubility results in unacceptable pharmacoki-netic properties, low bioavailability, and possible toxicity minimizing its therapeutic effect. In an effort to improve the problems mentioned above, there are many bio-materials – known as excipients – that are used to improve solubility and improve the bioactive molecule effective administration. We can mention the following cat-egories of such biomaterials:

- Nonionic surfactants like polyethoxylated castor oil, polysorbate 80, etc.
- Water-soluble organic solutes like ethanol, polyethylene glycol 400, polypropyl-ene glycol, cyclodextrins, phospholipids, etc.

The resulted pharmacotechnological forms are characterized as conventional and apparently they contribute in per os administration. Despite of the important contribution of biomaterials-excipients in therapeutics, we should not confuse their unwanted side effects during their administration as components of the final medicine.

Toxicity, their low ability to solubilize the bioactive molecule, and fast release rate after administration are crucial factors for their reevaluation. Also, many of the so-called compatible biomaterials-excipients present unwanted side effects, like inflammations, pain during injectable administration, kidney toxicity and/or neuro-toxicity. For example, polysorbate 80 and Cremophor EL^R present side effects like peripheral kidney toxicity and acute anaphylactic reactions. According to all information mentioned above, in relation to biomaterials-excipients used in a medicine development, and especially for drugs of high toxicity and specialty (e.g., cancer), the development of new innovative nanotechnological systems for bioactive molecule delivery to the damaged tissues is obligatory. Biocompatibility and biodegradability of biomaterials that nanosystems are comprised of, e.g., polymers, lipids, macromolecules, etc., are the basic characteristics that need to be evaluated before clinical trials and finally for public consumption. Other important characteristics of nanoparticles that will be used as innovative carriers of bioactive molecules are their physicochemical and thermodynamic properties [8] and should be tested with the appropriate analytical techniques/assays (see Chap. 2.3).

The ideal properties of a delivery nanosystem is the size and shape and structural control, the biocompatibility, the lack of toxicity, the precise structural and chemical modifications, its nanometric scale, the loading capacity, and the bioactive molecule release rate. Based on pharmacokinetics, the ideal delivery nanosystem includes structural surface macromolecules for tissue targeting. The effective nanoparticle system cellular adhesion, endocytosis, and intercellular transport can allow bioactive molecule or imaging factor transport to the cell cytoplasm or nucleus. Nanosystem most important advantages are the ability of size, structure, and shape control as well as the ability of multiple-function integration in a single nanodevice or nanosystem.

Therefore, a test of their colloidal properties in various dispersion media must be made with the aid of a great variety of physicochemical parameters, like chemical composition, concentration, and temperature. In the recent literature [26], a Modulatory Liposomal Controlled Release System (MLCRS) is described that combines liposomal and dendrimer technology where the anticancer drug doxorubicin is entrapped and its release rate was studied. The aim was to study the behavior of two different biomaterials in a nanosystem regarding doxorubicin's release rate. Recent bioactive molecule carriers of nanometer size offer the ability of increased therapeutic effectiveness, lowering the toxicity against normal tissues and achieving controlled therapeutic levels for extended time. Also, they can increase the lipophilic bioactive molecule solubility and increase their stability and allowing the completion of preclinical or clinical studies. Lastly, nanotechnology can help in the development of multifunctional systems with targeted bioactive molecule delivery and nanosystems with simultaneous diagnostic and therapeutic properties.

Despite of the fact that nanotechnology offers the ability of developing effective nanosystems for bioactive molecules [19] of all kinds of administration and delivery, the major orientations are toward nanosystem development for bioactive molecule administration in diseases of the respiratory system and the central nervous system. At the same time, nanosystems for administration and delivery of anticancer agents, hormones, and vaccines are being developed.

In a recent study, Rowland and coworkers [31] mention: "A drug delivery system is the pharmacotechnological system that carries bioactive molecules in the human body improving their efficacy, safety, controlling the release rate, time and choosing the tissue where the bioactive molecule will be released." According to our approach, the development process of delivery nanosystem development should be based on the following three important aspects: the biomaterial cooperativity in order to create a homogeneous behavior in the biological environment; the quality characteristics of the individual biomaterials that are related to their chemical structure and their physicochemical characteristics; and lastly, the nanosystem phase transitions, i.e., *metastable phases* (see Chap. 2.3), which should be studied. Even though the nanosystem physicochemical characteristics are sufficiently evaluated by the material science and science of colloidal dispersion systems, it should be taken under consideration that the nanosystem physicochemical behavior after in vivo administration should be reevaluated. Figure 5.1 presents important bioactive molecule delivery nanosystems.

The question arising during nanosystem design and development is whether these nanosystems can mimic the behavior or better the functionality of living cells. The science of interfacial phenomena – which have a definitive part through biological processes that are strictly defined and evolve according to adjustability potential of the living systems – requires nanosystem surface physicochemical parameter control through the science of biophysics and nanotechnology, in order to simulate the organism cell membrane functionality. This functionality is directly related to the liquid crystalline state (see Chap. 2). Also, the phase transitions (see Chap. 2.3) of membranes are thermodynamically *metastable phases* and contribute in the positive energy balance (energy accepted vs. energy offered), related to their functionality. These *metastable phases* should be definitely studied and simulated biophysically with the living cell membrane [8]. The characteristic and functions of a nanosystem designed to simulate cell's functionality are the self-assembly, the functionality according to the interfacial phenomena, and control over liquid crystalline phase transitions and its characteristics like size, shape, and physicochemical properties.

Bioactive molecule delivery nanosystem classification is not easy. Crommelin and Florence [6] classified them into the following:

• Targeting systems
• Biofeedback systems
• Remote control systems

The remarkable evolution in material sciences in the recent decades is the driving force to reconsider the term excipients that are used as classic carriers of bioactive molecules. It is obvious that in the near future new approaches and concepts will set

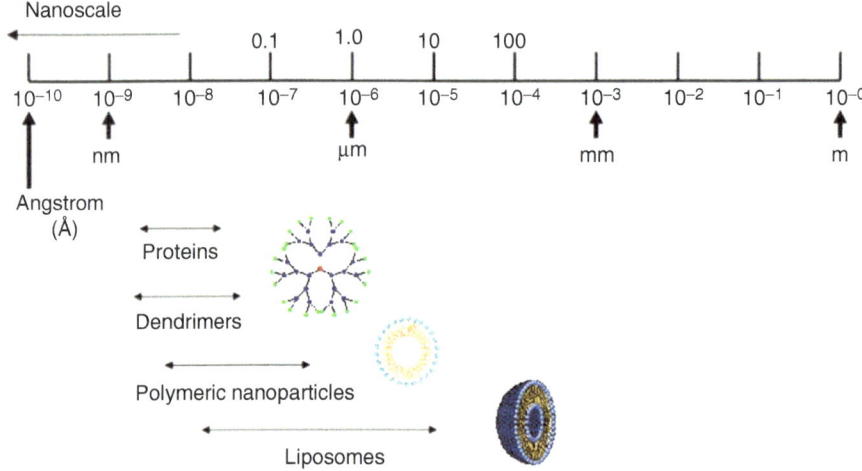

Fig. 5.1 Important drug delivery nanosystems

up regarding *innovative excipients* with self-assembled properties. The *functional excipients* are used characterizing biomaterials that improve the quality and effectiveness of the final medicinal product. However, nanoscaled drug delivery systems such as liposomes, dendrimers, polymersomes, etc. could be introduced as *self-assembled innovative excipients*. This approach could be the tool that nanosimilar drug delivery products will develop under an analytical new and rational concept (see Chap. 6) [10].

An effort to further classify drug delivery nanosystems resulted to a proposal published in *Drug Delivery* journal by Demetzos and Pippa [9]. The nature of the biomaterials being used is a critical factor for nanosystem classification. In the first case where nanosystems are of the same nature, i.e., lipids, the nanosystem is defined as *hybrid (Hy-)* [33], while in the second case where biomaterials are of different nature, i.e., lipids and polymers, the system is defined as *chimeric (Chi-)* from Chimera, the mythical form of the ancient Greek mythology [29]. The *metastable phase* determines the *hybrid* or *chimeric* membrane structure (Fig. 5.2). Examples of innovative drug delivery nanosystems are presented in Table 5.3.

The innovative drug delivery nanosystems classified as *hybrid* and *chimeric* consisting of the same or different biomaterials, respectively, should design, develop, and evaluate according to the following:

- Cooperativity between them. How much and in what way the individual biomolecular units develop homogeneous network that deliver information, having as a motivating force the chemical potential (μ) of the individual biomaterials
- The nature of biomaterials, i.e., polymer lipids, macromolecules, proteins, or even metabolism products of an organism
- The biomaterial phase transitions, in the final network of the drug delivery nanosystem according to its thermodynamics that is defined from the environment's physical parameters, its physicochemical characteristics, and its dynamic evolution in time

Fig. 5.2 Representation of the structure of liposomal membrane with anchored gradient block copolymer for creating *chimeric* nanosystems (Adapted from [29] with permission from Elsevier)

Table 5.3 Examples of innovative drug delivery nanosystems

System	Bioactive molecule	Biomaterials and/or inorganic materials	References
Liposomes in Liposomes (LiLs)	Leuprolide	Lipids	[33]
MLCRs/Chi-aDDnSs	Doxorubicin	Lipids/(newly synthesized) dendrimers	[12–14]
Chi-aDDnSs (targeted nanospheres and pretargeted radioimmunotherapy)	Paclitaxel	Polymers (ABA-type copolymers, PEG)	[4]
Chi-aDDnSs	Indomethacin	Lipids/gradient block copolymers	[29, 30]

These innovative drug delivery nanosystems according to the characteristics mentioned above, following the biologic system functionality, lead according to their surface composition toward the development of bio-inspired drug delivery nanosystems. The use of dendrimers incorporating or attached to liposomes during the past years is highly interesting (see Chap. 4) as modulators in anticancer drug release rate. Nanosystems composed of liposomes and dendrimers have been named as LLDs (Liposomal "Lock in" Dendrimers) [12] or by the current name *chimeric* drug delivery nanosystems (*Chi*-DDnSs) and are possibly the base of a new field of research that combines liposome and dendrimer technology and targeting the change of bioactive molecule bioavailability. The idea of LLDs or *Chi*-DDnSs came as the natural evolution of nanomaterial entrapment into liposomes. Figure 5.3 represents an LLD system. LLD technology consolidation took place in 2002 by Khopade and coworkers [18], for the chemotherapeutic bioactive agent, methotrexate, and increased entrapment percentage in liposomes. It has been previously mentioned from previous studies that the methotrexate entrapment can be increased by the use of positively charged lipid, stearylamine, by connecting the bioactive molecules in lipids in order to produce pre-liposomes. The problem with all these pharmacotechnological developments was the bioactive molecule, methotrexate, and low entrapment percentages. Taking into consideration that this specific bioactive molecule demands high dosage administration, Khopade and coworkers [18] used cationic dendrimers of PAMAM group to increase the methotrexate entrapment percentage into liposomes. This increase was due to the alkaline environment inside the liposomes that creates pH gradient. The entrapped dendrimer was assumed to act as

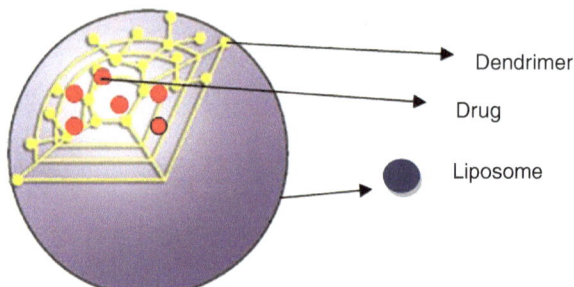

Dendrimer

Drug

Liposome

Fig. 5.3 Schematic representation of a LLD/chimeric liposomal drug delivery nanosystem (Chi-DDnSs) incorporated a drug

a tank for methotrexate flow inside the liposome while there was a decrease in the bioactive molecule release rate from the liposomal nanosystem.

PAMAM-doxorubicin liposomal nanosystem complex study was based on this observation [26]. The specific LLD allowed doxorubicin prolonged release from the liposomal nanosystem in relation to the conventional liposome, while it increased doxorubicin effectiveness in studies of in vitro breast and lung cancer series. By this successful technology application of dendrimers and liposomes, the term Modulatory Liposomal Controlled Release System (MLCRS) [26] was introduced in the field of nanotechnology regarding the controlled release liposome nanosystem use, whose evolution – according to our scientists' many years of efforts in the pharmaceutical nanotechnology laboratory in the National and Kapodistrian University of Athens – leads to the international literature consolidation of the term *chimeric* drug delivery nanosystems (*Chi*-DDnSs). In the specific study, the dendrimer molecule through the lipophilic complex with doxorubicin plays the modulator part.

Another LLD approach is the immobilization of the DPPG liposomes in layer-by-layer film with the aid of a PAMAM fourth-generation dendrimer. This method seems to be very promising for the bioactive molecule transdermal delivery and the biosensor development through a creation of porous films that could entrap ions and small molecules. Small molecule release from liposomes can take place through diffusion from dendrimer layers. The most recent LLD approach is related to bioactive molecule transport in cancer photodynamic therapy. This new system seems promising since the dendrimer photosensor can lead directly to the cancer tissue with the aid of the liposomal carrier. Despite of the inspiring results from the LLD technology, these hybrid systems seem to have stability issues since dendrimer presence causes lipid cyst agglomeration and fusion. In order to develop physicochemically stable LLDs with desired bioactive molecule entrapment and release characteristics, the physicochemical interactions between its component control and study are very important.

So far the studies have used PAMAM dendrimers with various end groups in their structure, and their interaction with the lipid membranes has been mostly studied with atomic force microscopy (AFM), ^{31}P NMP spectroscopy, and differential scanning calorimetry (DSC) techniques. The results related to PAMAM dendrimer interaction with lipid bilayers.

Recently, other polymers with similar architecture to dendrimers known as hyper-branched polymers (HBPs) have been used to develop *chimeric* drug delivery nanosystems [22]. Combining liposomes and amphiphilic block copolymers, we can secure more stable nanosystems also known as sterically stabilized nanosystems. Sterically stabilized liposomal nanosystems are the second-generation liposomes and are characterized of important properties including their high stability. Sterically stabilized liposomal nanosystem technology was achieved by modifying the liposomal membrane surface with conjugated hydrophilic polymer (like polyethylene glycol, PEG) engineering that is the most commonly used conjugated polymer, while at the same time a decrease intake from the immune systems unicellular phagocytes is observed, since the circulation time is increased with the PEG-phospholipid increased concentration.

Sterically stabilized liposomal nanosystems can be developed with two different ways, by grafting or by polymer absorption onto the liposomal surface. Phospholipid bilayer penetration from water-soluble surfactant block copolymers is an interesting approach for the bioactive molecule controlled release rate design and development. Stabilized nanosystem nontoxicity, high biocompatibility, and low immunogenicity and antigenicity are interesting aspects for clinical applications. Polymer-stabilized liposomes are usually used to transfer bioactive molecules targeting the prolonged release in bloodstream. It is generally accepted that the durability of these stereochemically protected nanosystems from the immune system macrophages is due to the suppressed opsonization process and to protein absorption to their surface that function as flags for their localization from the immune system macrophages. Innovative drug delivery nanosystems can be characterized as mixed nanosystems due to their combination of various biomaterials and are used for bioactive molecule delivery.

As mentioned, bioactive molecule controlled release nanosystems are classified as *hybrid* and *chimeric*. This nanosystem interest is due to the abilities they can offer for basic understanding of biological behavior, since biological systems use a range of mixed biomaterials to develop multifunctional self-assembly behavior.

5.2.1 Encapsulation and/or Incorporation of Bioactive Molecules into Nanosystems

Encapsulation (usually referred in water-soluble bioactive molecules) or incorporation (usually referred in lipophilic bioactive molecules) and their release rate from the nanosystem are important parameters for the nanosystem (that will carry the bioactive molecule) design and development.

The following definitions are related to the bioactive molecule encapsulation and/or incorporation in the nanosystem and are:

- Drug loading or drug content (DL) (%)

$$\text{Drug Loading } (\%) = \frac{\text{mass of bioactive molecule in nanosystem}}{\text{mass of nanosystem}} \times 100 \quad (5.1)$$

• Entrapment efficiency (EE) (%)

$$\text{Entrapment Efficiency } (\%) = \frac{\text{experimental bioactive molecule loading}}{\text{nominal biactive molecule loading}} \times 100 \quad (5.2)$$

It must be clarified that the percentage of the effective bioactive molecule encapsulation and/or loading describes the method effectiveness used for nanosystem's encapsulation and/or loading. The ideal method would achieve a great bioactive molecule encapsulation and/or loading, minimizing the nominal drug loading offering a great percentage of entrapment efficiency. It should also be mentioned that percentages should be referred as molecular ratios (fraction of bioactive molecule to the nanosystem) when the bioactive molecule nanosystem is homogenous to its composition, i.e., liposomes, composed only from lipids.

In case where the system is heterogeneous, i.e., lipids and polymers, in order to produce bioactive molecule transport nanosystems or better chimeric nanosystems, scientists must use the ratio of bioactive molecular mass to nanosystem ingredient mass to define the bioactive molecule encapsulation and/or loading into the nanosystem. Mostly the mechanisms referred to the bioactive molecule encapsulation and/or loading are the following: hydrogen bonds, ionic interactions, covalent bonds, and/or links and/or absorption onto the nanosystem surface.

The techniques used to study these interactions are nuclear magnetic resonance (NMR), ultraviolet spectroscopy (UV-vis), infrared spectroscopy (IR), and differential scanning calorimetry (DSC) or generally techniques for thermal analysis, X-rays, photoelectron spectroscopy (XPS), etc.

Bioactive molecule release from nanotechnological biomaterial matrix technology is the modern pharmaceutical research sector that is connected to physics as it takes advantage of the biomaterial physical properties to develop delivery nanosystems that affect the pharmacokinetics and therefore the pharmacological properties of the encapsulated or loaded bioactive molecule and are improved in comparison to the conventional pharmacotechnological formulations. The scientific culture of medicinal products nowadays more than any other time in history is formed from not only the science evolution but mostly of the technological evolution.

Release technology is the result of various scientific platforms, like pharmaceutics, polymer chemistry, and nanotechnology. Regarding the composition of the release nanosystem, it can be composed of biodegradable, nonbiodegradable, solubilized, biocompatible organic, or inorganic materials, while its form may vary from micro- and nanostructures to huge implants or membranes. In all carriers, a bioactive molecule diffusion phenomenon is implicated. The release mechanism is controlled from the diffusion phenomenon that is due to the escalated concentration gradient through polymeric network swelling or degradation (i.e., hydrolysis or biomaterial bond enzymatic degradation), in case of biodegradable polymeric biomaterials.

The two most commonly used terms to describe the mechanism for bioactive molecule release are:

- Sustained release
- Controlled release

The first mechanism is for pharmacotechnological systems that can sustain the bioactive molecule release and its circulation to the bloodstream to improve their therapeutic action. The term controlled release indicates transfer nanosystems with predictable and repeatable bioactive molecule release rate. During bioactive molecule administration, in order to obtain a pharmacological action, its concentration should slightly exceed its therapeutic concentration, while to avoid the toxicity symptoms, it should not exceed a maximum value that is referred as the maximum allowed concentration. The controlled release rate during application in pharmaceutical research presents important advantages but disadvantages as well.

The advantages refer to bioactive molecule concentration stabilization above the therapeutic levels for the treatment period, to less concentrations fluctuations, in lowering the side effects, in minimizing the bioactive molecule quantity needed for treatment, in reducing the dose administration frequency, in better patient compliance, and lastly in the effective bioactive molecule administration with small biological halftime life. On the other hand, the disadvantages refer to the high productivity costs, the possibility of toxicity due to materials used during production, the risk of bioactive molecule toxic levels due to entrapped bioactive molecule unexpected release, and possible patient's disturbance from the pharmacotechnological form or from the route of administration.

For many decades, the treatment of an acute or chronic disease was based on patients' drug administration in various pharmacotechnological forms like capsules, tablets, creams, solutions, and other conventional forms. These conventional bioactive molecule transport nanosystems are the pharmaceutical products predominant in the market.

Despite all these, to achieve and preserve the bioactive molecule concentration in the organism in the therapeutic levels, it is usually necessary to administer the dose many times during the day. As a consequence, an unwanted increase/decrease of bioactive molecule level in the bloodstream can result in decrease effectiveness of the dosage form and can be the cause of product low safety. At the same time, pharmaceutical products with controlled release technology have larger lifetime, and therefore this technology is more preferable in the pharmaceutical industry.

Modern technology allows bioactive molecule control release, extending the therapeutic action time (lowering the doses as well) or/and targeting in transport to a specific tissue (drug targeting). The interest has been moved to more complex pharmacotechnological forms of bioactive molecule administration for macromolecules like proteins, peptides, etc. that present difficulties when administered as pharmacotechnological formulations and are usually administered by injection. In addition, there is an effort to replace the macromolecular injection administration with other routes of administration, mostly per os, and to improve the quality of many biologically active molecules with greater safety due to dosage lowering. To achieve all the above, the interest is focused in research activities with very promising technologies like controlled drug delivery with innovative nanosystems.

5.3   Bio-inspired Drug Delivery Nanosystems

Nanosystems whose design follows the rules of living nature functionality are called bio-inspired drug delivery nanosystems and are characterized as innovative [21]. According to Carmen Alvarez-Lorenzo and Concheico [1], bio-inspired drug delivery systems are "The way nature design, processes and assembles molecular building blocks to fabricate high-performance materials with a minimum of resources is a suitable model for the design of drug delivery system (DDS) with advanced functionalities." According to this approach, they can be characterized as bio-inspired drug delivery nanosystems. These nanosystems are inspired from biological units like viruses, bacteria, cell structures, dendrimer structures, etc. that are present in living matter. The fact that these biosystems have the ability of self-assembly is due to the cooperativity of various materials they are composed of, e.g., lipids, phospholipid proteins, carbohydrates, nucleic acids, etc., which in an admirable cooperativity offers thermodynamic sufficiency for their stability, can achieve complex and multiple activities. The choice of different materials can lead to cooperativity that is unique to its result and can relate to nanosystem stability, entrapped bioactive molecule release rate control, and change of function mechanisms when conditions require continuous self-assembly to preserve its thermodynamic sufficiency. These bio-inspired systems are composed of different biomaterials that aim to achieve all actions mentioned before and can be characterized as *chimeric* bio-inspired drug delivery nanosystems (*chi*-aDDnSs) [30]. Bio-inspired drug delivery nanosystems that were mentioned above as well comprise the effort of biological system synthetic approach. The meeting area of synthetic and biological course of their production consists a new emerging scientific field of new delivery nanosystems for bioactive molecules and biomolecules like proteins, antibodies, macromolecules, etc. with biofunctionality similar to biological systems. Table 5.4 [37] presents biomimetic and bio-inspired nanosystems' categories and development strategies, applications, and uses. Bio-inspired nanosystems should follow the systems biology and pharmacology approaches.

5.4   Antibody-Drug Conjugates

Targeted drug delivery is a method of delivering bioactive molecules to a patient in a manner that decreases the side effects and increases the therapeutic index. The main goal of a targeted drug delivery system is to prolong, localize,

Table 5.4 Current development status of bioengineered, bio-inspired, and biomimetic drug delivery carriers

Strategies	Key attributes/capabilities	Applications	Current status	Challenges and/or limitations
Virus mimetics				
Recombinant bacteria	Tumor tropism No pathogenicity	Vaccine delivery	Clinical trials (phase I)	Safety concerns associated with attenuated bacteria (reversion to virulence)
Microbots	Carry nanoparticles on the surface of bacteria Neither bacterial disruption nor genetic manipulation is required Take advantage of the invasive property of bacteria	Gene or protein delivery	Preclinical	Safety concerns associated with attenuated bacteria (reversion to virulence) Applicability in actual disease models. Feasibility with biocompatible nanoparticles
Bacterial ghosts	No cytoplasmic contents. Intact surface properties Large drug-loading capacity Natural tropism to various tissues, including tumors Considerable safety and low production cost	Drug or DNA delivery Vaccine delivery	Preclinical	Potential immunogenicity owing to lipopolysaccharide Limited in vivo data
Viruses				
Viral vectors	Replace viral genetic materials with desirable ones Take advantage of transduction and self-replication ability of viruses Allow long-term expression of target genes. Carry nanoparticles	Gene therapy and/or imaging	Clinical trials (phase I and III)	Safety concerns (reversion to virulence). Limited targeting ability (off-target effects). Limited loading capacity
Viruslike particles	Self-assembled particles that are composed of viral capsids Easy to scale up at a low cost Preserve antigenicity Drug-loading capabilities Natural tropism and targeting ability with further modification	Vaccine delivery Drug and DNA delivery	Gardasil (Merck) και Cervarix (GlaxoSmithKline))	Potential immunogenicity when used for non-vaccine delivery

(continued)

Table 5.4 (continued)

Strategies	Key attributes/capabilities	Applications	Current status	Challenges and/or limitations
Virosomes	Reconstituted empty influenza virus envelope Easy to produce with low toxicity Adjuvant activity	Vaccine delivery Drug and DNA delivery	(Epaxal (Crucell), Invivac (Solvay influenza) και Inflexal V (Crucell)	Potential immunogenicity when used for non-vaccine delivery
Eukaryotic cells				
Red blood cells	Prolonged circulation (~120 days) Large volume for drug encapsulation Ability to carry nanoparticles and thrombolytics	Drug delivery	Preclinical	Difficult to maintain integrity Limited targeting ability
Macrophages	Natural homing tendency to disease sites Ability to move through the BBB Ability to phagocytose nanoparticles	"Trojan horse" delivery carriers	Preclinical	Difficult to collect Difficult to maintain integrity
Lymphocytes	Ability to carry various sizes of particles No damage to intrinsic functionality of the cells	"Cellular backpack" Adoptive T cell therapy of cancer	Preclinical	Difficult to collect Difficult to maintain integrity
Stem cells	Gene delivery by genetic engineering Natural homing tendency to solid tumors Ability to internalize nanoparticles	Cancer therapy	Preclinical	Difficult to collect Difficult to maintain integrity
Pathogen mimetic vaccines				
Pattern recognition mechanisms	Ability to stimulate immune cells using danger signals from pathogens via pattern recognition mechanisms Co-packaging of danger signals as adjuvants and antigens for improved immunization	Vaccine delivery	Preclinical	Limited to vaccine delivery
Virus mimetics				
pH-sensitive nanogels	Capsid-like structure pH-sensitive reversible swelling is followed by drug release and endosomal escape Ability to migrate to neighboring cells	Targeting tumors	In vitro	Vulnerable to immune recognition In vivo validation needed

Filomicelles	Flexible and filament-shaped micelles Prolonged circulation time in blood (over 1 week)	Targeting tumors	Preclinical	Thorough investigation into PK/PD needed
Cell mimetics				
Synthetic red blood cells	Ability to mimic shape and mechanical property of RBCs Drug-loading ability. Oxygen-carrying ability	Drug delivery Component of artificial blood	Preclinical	Vulnerable to immune recognition Detailed in vivo validation needed
Self-marker CD47	Membrane protein that is derived from RBC Contributes to self-recognition of RBCs by RES, thus enabling prolonged circulation time	Evasion of RES	In vitro	Limited resource
Compartmentalization				
Vesosomes	Liposomes within a liposome: distinct inner compartments separated from the external membrane Sustained release profile	Drug delivery	Preclinical	Vulnerable to immune recognition In vivo validation needed
Nanocells	Polymer nanoparticles within lipid vesicles Dual drug release system: rapid release of one drug from the lipid layer and sustained release of the other drug from polymer nanoparticles	Cancer therapy	Preclinical	Vulnerable to immune recognition

Adapted from [37] with permission from Nature Reviews Drug Discovery
BBB blood-brain barrier, *FDA* US Food and Drug Administration, *GRAS* generally regarded as safe, *PK/PD* pharmacokinetics/pharmacodynamics, *RBC* red blood cells, *RES* reticuloendothelial system, *siRNA* small interfering RNA

and target and have a protected drug interaction with the diseased tissue. The advantages to the targeted release system are the reduction in the frequency of the dosages taken by the patient, having a more uniform effect of the drug, reduction of drug side effects, and reduced fluctuation in circulating drug levels.

Antibody-drug conjugates (ADC) represent an innovative therapeutic application that combines the unique properties of monoclonal antibodies with the potent cell killing activity of cytotoxic bioactive compounds [25]. ADCs are complex molecules composed of an antibody linked, via a stable, chemical, linker with labile bonds, to a biological active cytotoxic (anticancer) payload or drug. The key components of ADC include a monoclonal antibody, a stable linker, and a cytotoxic agent to target a variety of cancers. Antibody-drug conjugates are single molecular species (or as close to single molecular species as current manufacturing can produce). It should be noted that antibody-drug conjugates are not nanoparticles but belong to molecularly targeted therapy. Monoclonal antibody therapy is a category of immunotherapy including activation/suppression immunotherapies that uses monoclonal antibodies to specifically bind to target (cells and/or proteins) [35].

On the other hand, according to Beck and Reichert [2], ADCs are a subclass of antibody-related therapeutics. Antibody-drug conjugates are a new class of highly potent biopharmaceutical medicines designed as a targeted therapy for the treatment of people mainly with cancer. ADCs are complex molecules composed of an antibody linked, via a stable, chemical, linker with labile bonds, to a biological active cytotoxic (anticancer) payload or drug [20]. By combining the unique targeting capabilities of monoclonal antibodies with the cancer-killing ability of cytotoxic drugs, antibody-drug conjugates allow sensitive discrimination between healthy and diseased tissue. This means that, in contrast to traditional chemotherapeutic agents, antibody-drug conjugates target and attack the cancer cell so that healthy cells are less severely affected [20]. ADCs deliver highly potent cytotoxic anticancer agents to cancer cells by joining them to monoclonal antibodies by biodegradable linkers and discriminate between cancer and normal tissue. The stability of ADCs is due to biodegradable linkers, which are either cleavable or non-cleavable. The advantages of ADCs are the increase of the cell-killing potential of monoclonal antibodies, the higher tumor selectivity, the increase of drug tolerability, and limited systemic exposure (compared to standard chemotherapeutic bioactive molecules) [32]. The only disadvantage of the ADCs is the difficulties of linker technology because the design/synthesis of a bio-functional is remarkably challenging due to the requirements of specialized teams [24]. Ado-trastuzumab emtasine and brentuximab vedotin are ADCs that are in clinical use (Figs. 5.4 and 5.5). These ADCs selectively deliver cargoes to tumor cells and provide clinical benefit by minimizing systemic toxicity. The mechanism of action and the biological activity are represented in Fig. 5.6.

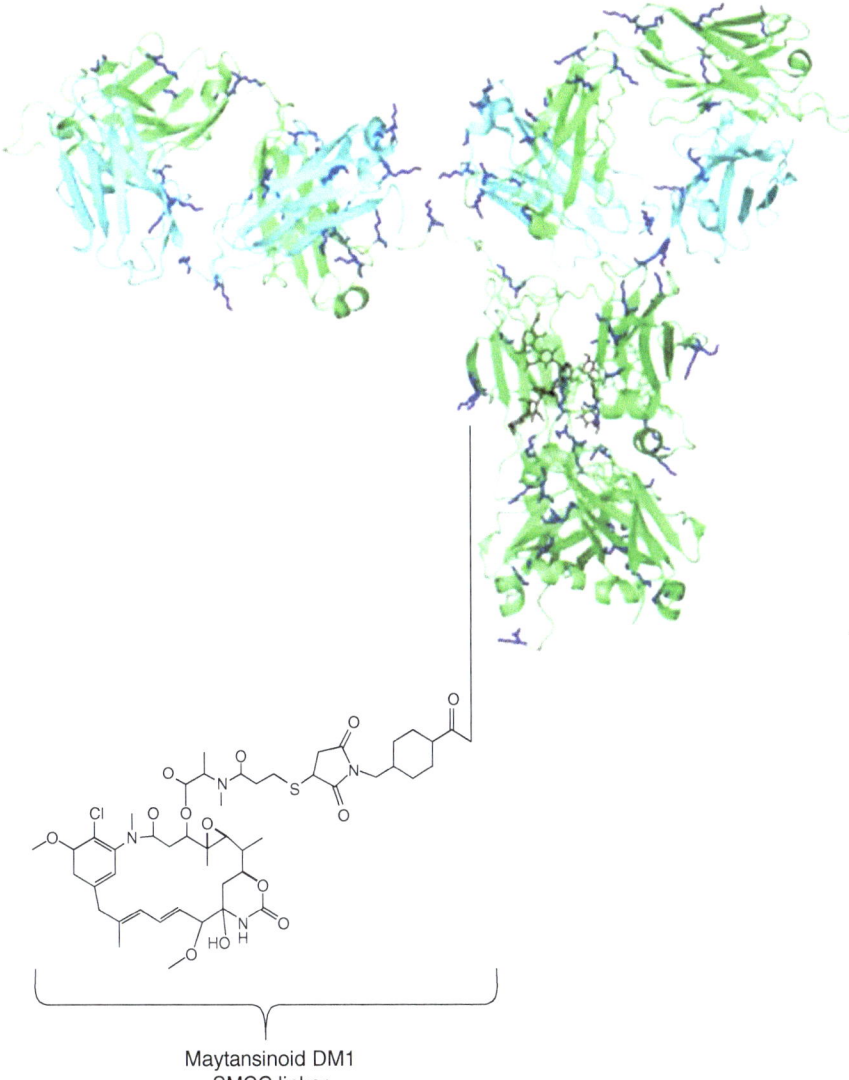

Maytansinoid DM1
SMCC linker

Fig. 5.4 Schematic representation of trastuzumab emtasine in which the drug maytansinoid DM1 is linked to lysine via thioether SMCC. The sticks represent more than eight naturally occurring lysines (Adapted from [11])

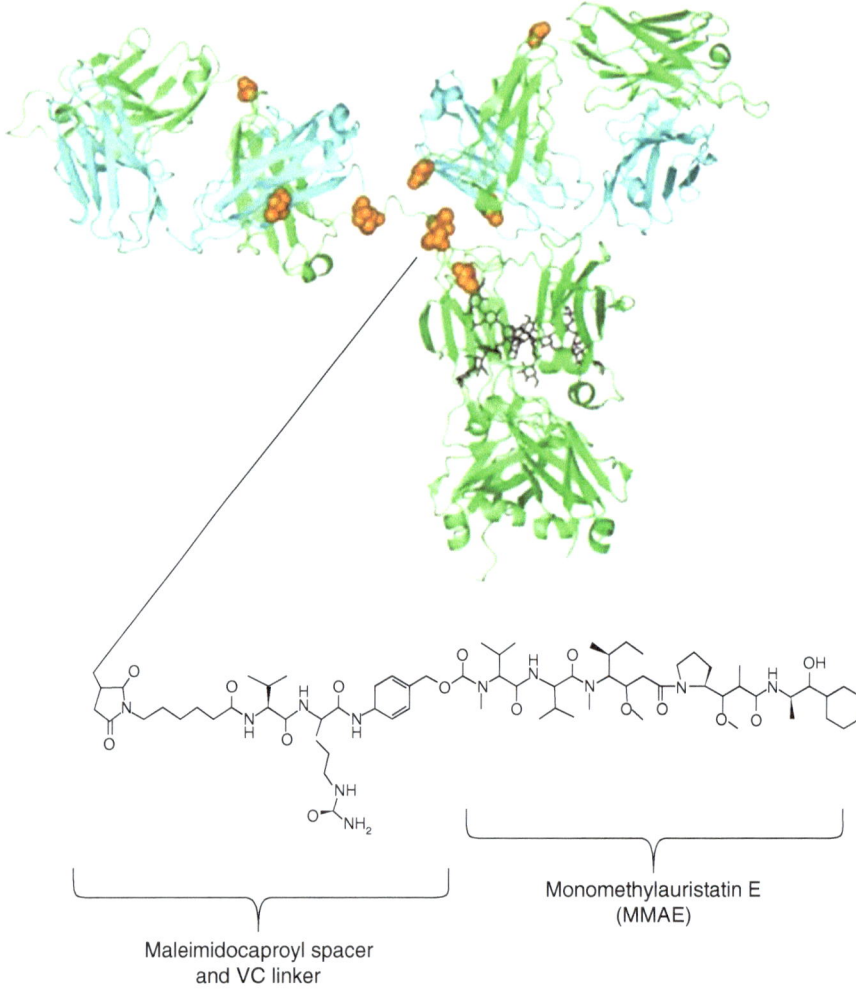

Fig. 5.5 Schematic representation of brentuximab vedotin which shows the conjugation of cysteines via maleimidocaproyl-VC dipeptide-PAB-MMAE. The spheres represent the eight naturally occurring cysteines (Adapted from [11])

Appendix

Thermodynamics of Innovative Drug Delivery Nanosystems

Calorimetry is a technique that can be used to measure the heat capacity of a material with a well-defined mass. It is also used to determine thermal effects that occur in physical, chemical, and biological processes. It is of interest that changes of

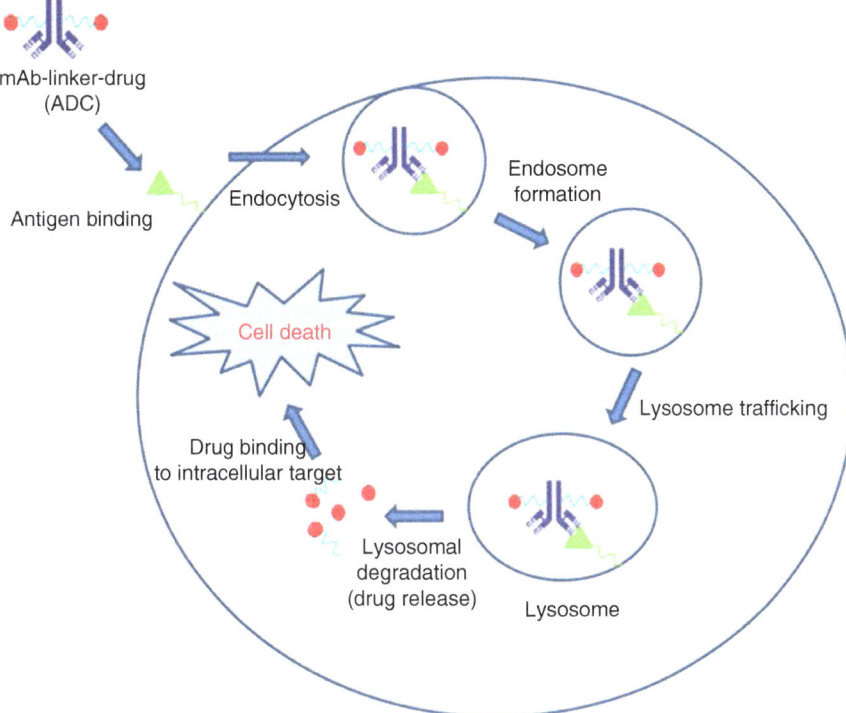

mAb-linker-drug
(ADC)

Antigen binding

Endocytosis

Endosome
formation

Cell death

Lysosome trafficking

Drug binding
to intracellular target

Lysosomal
degradation
(drug release)

Lysosome

Fig. 5.6 Schematic representation of the action mechanism of antibody-drug conjugates (Adapted from Bourchard et al. [3] with permission from Elsevier)

enthalpies in mixed materials such as biological objects or biomaterial-formed nanoparticles are issued to the field of calorimetry.

Thermodynamics is considered as a basic element in physics, and thermal analysis is used to determine the thermal effects of materials and biomaterials, and it is considered as one of the most popular techniques in material sciences and engineering [17]. There are a numerous thermoanalytical techniques (see Chap. 2) which can provide information such as polymorphism, stability, interactions, and physical purity of biomaterials or of self-assembled nanostructures such as liposomes, micelles, and nanoparticles (Gardikis et al. [13]). There are several applications that can be considered as very helpful in order to establish the behavior of biomaterials and of nanostructures. We can point out that processes such as melting, polymorphism, purity vaporization, glass transition, compatibility between different biomaterials in nature, etc. are highly related to the field of thermodynamics. Supramolecular structures at nanoscale level are very promising and can be used in drug delivery in imaging and in diagnosis. The biophysics and thermodynamics are considered as scientific building blocks for evaluating drug delivery systems [8]. Biophysics is a valuable element in order to explain the orientation and reorientation process of a nanosystem taking into account the external environmental parameters, and it is also used to understand the organization of nanostructures comparing their behavior with that of living organisms. It is well established that different phases such as *cubic*, *hexagonal*, *inverse hexagonal*, etc. can be produced depending on the physical *micro-* and *macro*-environment parameters such as temperature, ionic strength, pH, osmolarity, and concentration of lipids to be used for producing liposomal bilayers. The aforementioned changes can be directly correlated with physicochemical and biophysical behavior of liposomes as drug delivery systems, such as kinetics in drug release, stability, surface functionality, etc.

The melting process of lipidic drug delivery nanosystems, i.e., liposomes, affects the behavior of lipidic membranes and induces events leading to transitions and creation of *metastable phases* (see Chap. 2). Thermodynamics is considered as a basic element in physics, and thermal analysis is used to determine the thermal effects of materials and biomaterials, and it is considered as one of the most popular techniques in material sciences and engineering. There are a numerous thermo-analytical techniques (see Chap. 2) which can provide information such as polymorphism, stability, interactions, and physical purity of biomaterials or of self-assembled nanostructures. The naturally occurring infrastructures are composed of different nature biomaterials such as hydrocarbons, proteins, lipids, etc., and this could be an efficient model that can promote artificial structures and strategies in order to develop innovative drug delivery nanosystems. By combining different nature biomaterials such as lipid and polymers, *chimeric* artificial supramolecular structures can be built up (see Chap. 5) [14]. Their thermodynamic profile contributes to the development process and to study their equilibrium states by passing from one *metastable phase* to another. The steady state is almost eliminated and their organization profile promotes their functionality and can be related with their stability. A Gibbs free energy difference, Δ (ΔG), is an appropriate thermodynamic parameter to initiate the equilibrium state in innovative mixed in nature bilayer between different membrane segments of different curvatures. This attributes to a high involvement of thermodynamics in studying and measuring the transitions between lipidic phases. The formation of lipidic vesicles has been extensively studied by researchers mainly in the field of colloidal science, and the polymorphism of phospholipids seems to play a key role because of its relation with the geometrical characteristics of liposomal vesicles. The geometrical aspects and the thermodynamics and their compromise relationship are under investigation especially with those vesicles that are categorized as *chimeric* drug delivery nanosystems (Chap. 5). The *metastable phases* are considered as important element in order to realize the thermal behavior of lipidic bilayers. The lipid-water systems can create lipidic phases with predominant *lamellar* phases, depending on the lipid concentration. Liposomes belong to drug delivery nanosystems and are characterized as lyotropic liquid crystals (see Chap. 4). The conformational polymorphism of their lipidic bilayers is responsible for the *mesophases*, i.e., *metastable phases*, taking place in phase transitions and is related to their thermal stress during phase transitions. Their thermal stress takes place during liposome dispersion system storage or during administration in humans. The thermodynamic parameters that affect and participate in physical stability and, therefore, in pharmaceutical effectiveness of the liposomal product are as follows: T_m, temperature of basic transition from liquid crystalline phase to isotropic fluid; $\Delta T_{1/2}$, the phase transition range in the middle of the peak (this temperature range is related to the cooperativity of system phospholipids or phospholipids and enclosed bioactive molecule); ΔH, system enthalpy change; and ΔC_p, system thermal capacity change under constant pressure. The identification and study of these *mesophases* of liposomal nanosystem lipid bilayers allow the control over the thermodynamic parameters mentioned above, in order to rationally design the liposomal system with the most satisfactory physical and thermal stability [7].

Thermal techniques are considered as valuable to study the thermal behavior of the conformational polymorphism (i.e., *metastable phases*) of drug delivery nanosystems, such as liposomes, while they are used to evaluate the physicochemical properties of drugs and their interactions in in vitro biological media, as well as their behavior during the formulation process [28]. The regulatory issues concerning the design and the development of innovative drug delivery nanosystems involve a combination of valuable techniques for structural and thermodynamic characterization in order to completely delineate the physicochemical/thermodynamic balance [28].

From the regulatory point of view, thermal analysis can provide rational approaches to fully characterize delivery nanosystems that could be applied in pharmaceutics, adopting the thermodynamics in the evaluation of the development process of medicines.

Summary

Innovative drug delivery nanosystems are classified as *hybrid* and as *chimeric* in order to facilitate their studies based on biophysical and thermodynamic aspects. The above terms are the evolution of the Liposomal *Lock in* Dendrimers (LLDs).

The *metastable phases* of *hybrid* and *chimeric* nanoparticulate systems determine their stability, release of the incorporated bioactive molecule, and functionality and can translate the cooperativity of their building elements and their biophysical behavior based on the origin of the biomaterials that they are composed of.

The innovation profile of bio-inspired drug delivery nanosystems is based on their ability to mimic the function of natural objects. Moreover, their fabrication process could adapt various aspects that nature use to build up high-performance biological units.

References

1. Alvarez-Lorenzo C, Conheiro A (2013) Bio-inspired drug delivery systems. Curr Opin Biotechnol 24:1–7
2. Beck A, Reichert JM (2014) Antibody-drug conjugates: present and future. MAbs 6(1):15–17
3. Bouchard H, Viskov C, Garcia-Echeverria C (2014) Antibody-drug conjugates—a new wave of cancer drugs. Bioorg Med Chem Lett 24(23):5357–5363. doi:10.1016/j.bmcl.2014.10.021.

4. Bushman J, Vaugham A, Sheihet L et al (2013) Functionalized nanospheres for targeted delivery of paclitaxel. J Control Release 171(3):315–321
5. Chang TMS (1979) Artificial cells as drug carriers in biology and medicine. In: Gregoriadis G (ed) Drug carriers in biology and medicine. Academic, London, pp 271–285
6. Crommelin DJ, FFlorence AT (2013) Towards more effective advanced drug delivery systems. Int J Pharm 454:496–511
7. Demetzos C (2008) Differential scanning calorimetry (DSC): a tool to study the thermal behavior of lipid bilayers and liposomal stability. J Liposome Res 18:159–173
8. Demetzos C (2015) Biophysics and thermodynamics: the scientific blocks of bio-inspired drug delivery nano systems. AAPS PharmSciTech 16(3):491–495
9. Demetzos C, Pippa N (2014) Advanced drug delivery nanosystems (aDDnSs): a mini review. Drug Deliv 21(4):250–257
10. Emshanova SE (2008) Drug synthesis methods and production technologies, methodological approaches to the selection of excipients for preparation tablets by direct pressing. Pharm Chem J 42(2):89–94
11. Feng Y, Zhu Z, Chen W et al (2014) Conjugates of small molecule drugs with antibodies and other proteins. Biomedicines 2:1–13. doi:10.3390/biomedicines2010001
12. Gardikis K, Hatziantoniou S, Bucos M et al (2010) New drug delivery nanosystem combining liposomal and dendrimeric technology (*liposomal-locked in dendrimers*) for cancer therapy. J Pharm Sci 99(8):3561–3571
13. Gardikis K, Hatziantoniou S, Signorelli M et al (2010) Thermodynamics and structural characterization of liposomal locked-in dendrimers as drug carriers. Colloids Surf B Biointerfaces 81(1):11–19
14. Gardikis K, Tsimplouli C, Dimas K et al (2010) New chimeric advanced drug delivery nanosystems (Chi-aDDnSs) as doxorubicin carriers. Int J Pharm 402(1–2):231–237
15. Gregoriadis G (2008) Liposome research in drug delivery, the early days. J Drug Target 16(7):520–524
16. Gupta H, Bhandari D, Sharma A (2009) Recent trends in oral drug delivery: a review. Recent Patents Drug Deliv Formulation 3(2):162–173
17. Heimburg T (2007) Thermal biophysics of membranes. Wiley –Vott, Weinheim
18. Khopade AJ, Caruso F, Tzipathi P, Nagaich S, Jain NK (2002) Effect of dendrimer on entrapment and release of bioactive from liposome. Int J Pharm 232(1–2):157–162
19. Kiparissides C, Kammona O (2008) Nanotechnology advances in controlled drug delivery systems. Phys Stat Sol 5(12):3828–3833
20. Lianos GD, Vlachos K, Zoras O et al (2014) Potential of antibody-drug conjugates and novel therapeutics in breast câncer management. Onco Targets Ther 7:491–500
21. Lin Y, Mao C (2011) Bio-inspired supramolecular self-assembly towards soft nanomaterials. Front Matter Sci 5(3):247–256
22. Mourelatou EA, Libster D, Nir I et al (2011) Type location and interaction between hyperbranched polymers and liposomes. Relevance to design of potentially advanced drug delivery nanosystem (aDDnSs). J Phys Chem B 115(13):3400–3408
23. O' Neil GJ (1979) The use of antibodies as drug carriers. In: Gregoriadis G (ed) Drug carriers in biology and medicine. Academic, London, pp 23–41
24. Ornes S (2013) Antibody-drug conjugates. Proc Natl Acad Sci U S A 110(34):13695
25. Papachristos A, Pippa N, Demetzos C et al (2015) Antibody-drug conjugates: a mini-review. The synopsis of two approved medicines. Drug Deliv, in press
26. Papagiannaros A, Dimas K, Papaionannou GT, Demetzos C (2005) Doxorubicin-PAMAM dendrimer complex attached to liposomes: cytotoxic studies against human cancer cell lines. Int J Pharm 302:29–38
27. Peppas NA (2013) Historical perspective on advanced drug delivery: how engineering design and mathematical modeling helped the field nature. Adv Drug Deliv Rev 65(1):5–9
28. Pippa N, Gardikis K, Pispas S et al (2014) The physicochemical/thermodynamic balance of advanced drug liposomal delivery systems. J Therm Anal Calorim 116:99–105

29. Pippa N, Merkouraki M, Pispas S et al (2013) DPPC:MPOx chimeric advanced drug delivery nanosystems (chi-aDDnSs): physicochemical and structural characterization, stability and drug release studies. Int J Pharm 450(1–2):1–10
30. Pippa N, Kaditi E, Pispas S et al (2013) PEO-b-PCL: DPPC chimeric nanocarriers: self – assembly aspects in aqueous and biological media and drug incorporation. Soft Matter 9:4073–4082
31. Rowland M, Noe CR, Smith DA et al (2012) Impact of the pharmaceutical sciences on health care: a reflection over the past 50 years. J Pharm Sci 101:4075–4099
32. Sapra P, Betts A, Boni J (2013) Preclinical and clinical pharmacokinetic/pharmacodynamic considerations for antibody-drug conjugates. Expert Rev Clin Pharmacol 6(5):541–554
33. Saroglou V, Hatziantoniou S, Smyrniotakis M et al (2006) Synthesis liposomal formulation and thermal effects on phospholipid bilayers of leuprolide. J Peptide Sci 12(1):43–50
34. Stazz C (2000) Innovation in drug delivery. Patent Care 15:107–137
35. Shefet-Carasso L, Benhar I (2014) Antibody-targeted drugs and drug resistance-challenges and solutions. Drug Resist Updat 18:36–46
36. Tiwari G, Tiwari R, Sriwastawa B et al (2012) Drug delivery systems: an updated review. Int J Pharm 2(1):2–11
37. Yoo JW, Irvine JD, Discher DE et al (2011) Bio-inspired, bioengineered and biomimetic drug delivery carriers. Nat Rev Drug Discov 10:521–535

Chapter 6
Nanotoxicity and Biotoxicity

Abstract The toxicity of nanoparticles seems to emerge as the main concern in the near future. This concern regards with their future reproductive possibility through self-assembly mechanisms and of their stability that may pose a risk to human health. It is worth to note that their toxicity could be associated with their physico-chemical properties like size, size distribution, charge, and surface properties. Moreover their toxicity seems to be a result of their greater total area in comparison to particles of the same mass but larger dimensions. The toxicity in molecular level is defined as biotoxicity, while the route of administration seems to be essential for their accumulation in specific organs. Their clearance from the human organism occurs via the spleen and liver. The Food and Drug Administration (FDA), National Institute Of Safety and Health (NIOSH), and Environmental Protection Service realized that they should turn their attention in the nanosystem risk investigation so the appropriate safety rules and regulatory framework should be developed. However, vigilance in the approval of nanotechnological products was an emerging issue, and the manufactures and the regulatory agencies should act appropriately.

Keywords Fat and sticky fingers • Biotoxicity • Nanotoxicity • Nanosystem risk investigation • Environmental Protection Service

6.1 Toxicity and Safety of Nanosystems

6.1.1 Toxicity and Safety of Nanotechnological Products

Regarding the safety of human health from the use of nanotechnological products (nanosystems and nanodevices), there are objections and controversies. These controversies function in a positive way and are beneficial – when they do not use methods that lead to rejection of scientific product application in everyday life [16]. It is obvious that particle size reduction leads to better and greater permeability to human tissues, resulting in passage to systemic circulation and possibly selectively accumulation in organs and tissues. This is an initial evaluation that has references and citations.

© Springer Science+Business Media Singapore 2016   175
C. Demetzos, *Pharmaceutical Nanotechnology*,
DOI 10.1007/978-981-10-0791-0_6

The great concern does not regard the conventional nanostructure toxicity but their future reproductive possibility through self-assembly mechanisms and developing nanosystems of great stability that may pose a risk to human health. Richard Smalley, Nobel prized (1996) who discovered fullerene (carbon nanostructures), is reassuring and considers that this is possible to happen. Barriers closely related to manipulation of nanomaterials that prevent the construction and the development of reproducible nanostructures are the following:

- Fat fingers
- Sticky or colloidal fingers

Nanostructure development requires interaction between atoms not only neighbor atoms but distant ones as well [4]. The production of such nanostructures requires robotic nature high technology application that based on technology new achievements production rates is extremely distant. This is due to the phenomena mentioned above (i.e., fat and sticky or colloidal fingers). Nanostructure and nanosystem technological produce large fingers would be much larger – huge in relation to the person that handles production of self-assembly nanostructures/nanosystems. Also, robotic system towing bracket interactions would be powerful and would function adhesively restricting a number of movements.

We must also mention that nanostructure reproducibility requires massive amounts of constructive energy that will create quantum escape channels for these structures to other dimensions. If these nanostructures were created, their energy content would destroy them before their development. The future possibility of energy storage development during in situ nanomaterial self-assembly would make our planet energy sufficient and space station visit possible in minimum time.

Going back to nanosystem toxicity, the questions are many and related to safety of society health and personnel working in the field of health. Answers are vague and absolute safety data do not exist. The reports relate to everyday used nanotechnological products, and there is a possibility that the development and production processes do not follow the rules of good manufacturing practice, transport, and storage. Users of cleaning products that have been presented with health problems due to nanoparticles, the presence of which is debatable. It is obvious that nanoparticles' physicochemical characteristics, production, and development are not an easy manufacture.

High technology, qualified personnel, and most basically regulatory framework controlling the products production, transport, and storage are necessary. Therefore, we can say that nanosystem-proven risk through completed study does not exist, but this does not mean that we should stop looking for any hazards. Despite all these, studies in test animals show nanosystem possibility in entering human body faster than greater dimension particles. Their possible toxicity seems to be a result of their greater total area in comparison to particles of the same mass but larger dimensions. The easiest way to enter nanosystems into the organism is through the respiratory tract, while the depth of their penetration depends on their aerodynamics. Studies in test animals have shown that through respiration nanosystems can get into blood circulation and affect living organs. Another possible nanosystem administration route is through the gastrointestinal tract. The transfer could be achieved either per os or through the nozzles and reach the brain [5]. Titanium oxide nanoparticles were studied in Great Britain, and the results showed that they could penetrate the upper part of the skin (epidermis) without deeper penetration to the tissues [10,14,15]. The same results were presented in carbon nanotube applications. However, these results could not be considered equivalent to the ones on real conditions and not experimental/lab conditions.

Various research groups, like the one of Gunter Oberdorster in Rochester University, showed carbon nanosystems of 30–35 nm transport to the brain through the nozzles. Researchers in Duke University placed fish in water with a minimum carbon nanoparticle concentration (1 part nanosystem to 1,000,000 parts of water) and observed greater fish brain lipids in comparison to those not exposed, while in the research facility Johnson Space Center, NASA in Houston observed lung damages of mice exposed to carbon nanotubes.

Vigilance in the approval of nanotechnological products was directly decided as imperative, and in the United States of America, Food and Drug Administration (FDA), National Institute Of Safety and Health (NIOSH), and Environmental Protection Service realized that they should turn their attention in the nanosystem risk investigation so the appropriate safety rules and regulatory framework should be developed. Special attention should be taken for nanotechnology production workers according to NIOSH guidelines. This specific guideline includes guidance for possible ways that human organism and the precaution measures should be taken. The suggestions for employees' safety do not differ from the safety and hygiene rules that should be complied in labs and working places. The questions for nanosystem risk control rely on their evaluation parameters. Size, size distribution, composition, surface, charges, shape, surface properties, and stability in certain conditions are possible evaluation parameters.

At this point, it should be mentioned that toxicity kinetics of particles and nanoscale aggregations differ significantly. The parameters mentioned should be the *Red Book* for customers', employees', and researchers' protection in the nanotechnology field. Also, the development of parameter evaluation methods is an important challenge, but the nanostructure and nanodevice production processes in industrial scale are next decade challenges regarding their toxicology studies. Nanosystem, nanostructure, and nanodevice effect on the environment is an important challenge as well.

The biodegradability, the exposure, and the clearance kinetics of nanosystems require computer modeling in order to develop monitoring tools for safety of living organisms and environment. Nanotechnology product industry investments in the field of research and safety tool development in all stages of production and when in the market should be a primary objective. Research centers and universities should invest in nanosystem toxicity research and evaluate with scientific criteria the parameters that should be controlled during production, storage, and use as well as their behavior toward the environment.

Health products designed by nanotechnology and currently in the market include medicines, diagnostic, imaging, and cosmetic. The production of these products takes place under strict safety rules, and complex approval processes are followed from the National and International Approval Mechanisms. For the United States, there is the FDA and for Europe the European Medicines Agency (EMA) which demand strict specifications and completed submissions of paperwork before approval, while the production unit control follows more strict criteria.

We should mention that biomaterials being used and mentioned in the previous chapters (see Chaps. 2 and 4), e.g., bioactive molecule liposomal carrier production that are lipids, are biocompatible to human organism since they are cell membrane components, while materials like polymers are checked for toxicity before their production and bioactive molecule polymer carrier production.

Surely, we cannot claim that there is a reason for concern for nanosystem use that is present in everyday life. There should be a full risk assessment record and evaluation so that the coordinated actions between research institutes, industry, and commonwealth develop guidelines to evaluate risk parameters that should be minimized.

It is known that the energy sources that we currently use, for example, are extremely dangerous for our health and for the environment without this prohibiting their use. Epidemiology studies have shown important dangers from micrometer particle air pollution that affects the respiratory and cardiovascular system, even sometimes leading to death. The risk degree for the respiratory system in relation to the air pollution is known as *London fog episode* (1952). It is also known that thousands of pharmaceutical substances and bioactive molecules are used for the production of new medicines with dangerous side effects that cause unwanted effects and even lead to death of hundreds of people. The risk benefit ratio is the one that will prove if the final product is useful, ensuring the restrictions from toxic and unwanted side effects [3, 6, 17].

Nanotoxicology science is a challenge and a new direction to the toxicology field. Responsible use of new technology will define the final benefit and lead to new research for their improvement. The environment effect research for nanotechnology products is a subject of research activity in the United States but in also in Europe and Japan. The studies include reliable risk and safety evaluation development for nanotechnology products in the environment.

The evaluation and approval processes for nanotechnological products follow the processes described in Chap. 7, since there are no specific guidelines for the evaluation of nanopharmaceutical products. The Center for Technology Assessment (CTA) is an organism related to health and environment and considers nanoparticles as hazardous. This hazard is related with nanosystem ability to penetrate the skin and from the respiratory system to reach tissues and organs without specifying neither their half-life nor their clearance time.

The International Committee on Harmonization (ICH) for the preclinical toxicity studies suggests and group controls presumed necessary after the use of pharmaceutical products (ICH, S6, 1997). These include genetic, immunological, and cancer tests. The biomaterials used as excipients, i.e., pharmacologically inactive biomaterials that participate in the therapeutic action of the final medicine, are being mentioned in toxicology studies in FDA website in the Inactive Ingredient Guide (IIG). It is obvious that these biomaterials are in micrometer-size scale (10^{-6} m). The question is what happens with biomaterials that are used as excipients and are in nanodimension scale. This question also arises in the *Handbook of Pharmaceutical Excipients* (HPE). Also the United Kingdom's Royal Pharmaceutical Society suggests that excipients of nanoscale dimensions should be evaluated for safety before final product development. So, the full nanosystem characterization physico-chemically and biologically should be a part of the final product evaluation. Also, the National Academy of Science (NAS), summarizing all the above, mentions that nanosystems and pharmaceutical nanoproduct evaluating devices should be tested and have the ability of detecting all parameters mentioned. It seems according to up-to-date experience that in the future, there will be a requirement for nanosystem physicochemical parameter test devices' development that will have increased detection abilities for nanomaterial properties and will be a demanding regulatory procedure before leading to the final decision for the product approval.

We must mention that the nanosystem physicochemical properties are very important for its interaction with organism cells in nanotoxicology level if in pharmaceutical nanotechnology level this is a prerequisite for technological formulation/development. Specific examples will be mentioned later on.

Another important phenomenon affecting nanosystem toxicity is the *differential absorption phenomenon*, closely related to unwanted/toxic effects of nanoscale particles [12]. This approach implies that nanoparticle intravenous administration is directed to their destination (target tissue) as long as plasma proteins are absorbed onto them (e.g., albumin and immunoglobulins) [12]. Recent research studies presented results from nanosystem surface coating with lipoprotein E that eased nano-system penetration to the brain by passing the blood-brain barrier. This phenomenon called *concept of differential absorption* can cause side effects and toxic results from the administered nanotechnological or biotechnological form. The *concept of differential absorption* precedes nanosystem interaction with cells and defines nanosystem presence in specific tissues or organs. The phenomena mentioned before are present regardless if nanoparticles come from natural or human sources (Table 6.1). Moreover, particle size and surface chemistry are related to alveolar capillary spittle (Table 6.2). At this point, we must mention that the respiratory system, due to its structure, is the system where most of the unwanted side effects and toxic actions related to nanoparticles are present. After inhalation, nanoparticles set on the respiratory system (nose, pharynx, lungs). The lungs are composed of airways that import/export air and of alveoli where there is air exchange [9]. Human lungs have a lining of 75–140 m^2 and 300 millions of alveoli. *Pseudo-spherical* nanoparticles with a diameter of <10 μm reach the alveoli, while larger ones set on the upper parts of the respiratory system. It is very difficult – even if impossible – to remove nanoparticles from the alveoli with medicinal methods [9].

6.2 Nanotoxicology: The Toxicology of Nanosystems

Commission Recommendation 2011/696/EU, EC 2011 *defines nanomaterial as any particulate substance with at least one dimension in the size range between 1 and 100 nm*. Their properties are characterized as specific, contrary to the larger in size substances. The nanosystems have in a unique way an increased surface area that in combination with their chemical structure and organization offers a huge potential

Table 6.1 Ultrafine particles and nanoparticles (<100 nm)

| Natural | Anthropogenic | |
	Unintentional	Intentional
Gas to particle conversions	Internal combustion engines	Engineered nanoparticles (controlled size and shape, designed for functionality)
Forest fires	Power plants	Metals, semiconductors, metal oxides, quantum dots/rods, fullerenes, nanotubes, nanowires, nanoshells, nanorings, etc.
Volcanos	Incinerators	Nanotechnology applied to many products: cosmetics, medicals, fabrics, electronics, optics, displays, etc.)
Viruses	Airplane jets	
Biogenic magnetite: (magnetotactic bacteria, protists, mollusks, arthropods, fish, birds, human brain, meteorite)	Metal fumes (smelting, welding, etc.) polymer fumes	
	Other fumes	
Ferritin (12.5 nm)	Heated surfaces, frying, boiling, grilling	
Microparticles (<100 nm)	Electric motors	
Activated cells		

Adapted from Buzea et al. [1] with permission from Springer
Natural and anthropogenic sources

Table 6.2 Particle size and surface chemistry-related alveolar capillary translocation

Particle size (nm)	Type	Translocation	Localization/effect
5–20	Gold nanoparticles, albumin coated	Performed	Via caveolae
8	Gold nanoparticles, albumin coated	Performed	Via "vesicles"
8	Gold nanoparticles, albumin coated	Performed[a]	Via caveolae
18	Iridium nanoparticles	Performed	Extrapulmonary organs
30	Gold nanoparticles	Performed[b]	Platelet
35	Carbon nanoparticles	Performed	Liver
60	Polystyrene nanoparticles, positive charge	Unknown if it happens	Thrombus
60	Polystyrene nanoparticles, negative charge	Unknown if it happens	No thrombus
80	Iridium nanoparticles	Carried out	Extrapulmonary organs
240	Polystyrene nanoparticles, lecithin coated	Carried out	Monocyte
240	Polystyrene nanoparticles, without coating	Not performed	No thrombus
400	Polystyrene nanoparticles	Not performed	

Adapted from Buzea et al. [1] with permission from Springer
Surface coating (chemistry) charge, size govern translocation
[a]Minimal
[b]Indirect evidence

for applications and the possibility for toxic effects in human body that is not obvious in the macroscopic matter. Some studies lead us to be suspicious that nanosystems can act toxically in living organisms or show a high degree of uncertainty in predicting their toxicity. This second characteristic is very important regarding their chronic toxicity. Human organism does not have those effective mechanisms to remove nanoparticles, for example, from the gastrointestinal system and the rest of the organs, in comparison to larger particles of microscopic and macroscopic scale. Nanoparticles can move very easily from the point of entrance, following the systemic circulation in other parts of the human body, even in the heart and brain.

Scientific knowledge is limited concerning nanosystem various effects. In the following, there are some illustrative effects:

- Long-term nanosystem accumulation inside the human organism (chronic toxicity)
- Per os nanosystem administration and nanosystem toxic effects from per os administration in human organisms
- Possible effects in reproduction
- Teratogenic effects
- Nanosystem transport to breastfeeding milk
- The relation between nanosystem physicochemical properties and their toxicity
- Nanosystem toxicity when used as nutrient additive in foods and cosmetics
- Nanosystem biological behavior (distribution, metabolism, clearance, toxicity in specific organs and tissues toxicokinetics)
- Nanosystem accumulation during digestion along the gastrointestinal tract

The toxicity of nanoparticles includes assays to determine cytotoxicity, acute toxicity, irritation activity, genotoxicity, repeated dose toxicity, carcinogenicity, and reproductive and developmental toxicity among others (Guidance on the Determination of Potential Health Effects of Nanomaterials Used in Medical Devices, published in 2015 by the Scientific Committee on Emerging and Newly Identified Health Risks (SCENIHR)).

6.3 Biotoxicity: The Nanotoxicity in Molecular Level

Apoptosis is a programmed cell death which takes place in all multicellular organisms due to biochemical processes and paths triggered and lead to cellular form changes and finally to death. Cell death can also be caused from an outside factor. This outer factor can cause cell damage through mechanical press or through exposition to toxic compounds. Obviously nanomaterials that we use in pharmaceutical nanotechnology and nanomedicine level should have the minimum degree of toxicity, so nanodevices and nanoparticles can pass through the immune system without being recognized, i.e., have stealth properties. Iron oxide (magnetite and maghemite) nanoparticles and liposomes are biocompatible and, therefore, have no toxic effects. An iron ion functional nanoparticle schematic representation is shown in Fig. 6.1 (where iron magnetic nucleus is connected with polymer shell

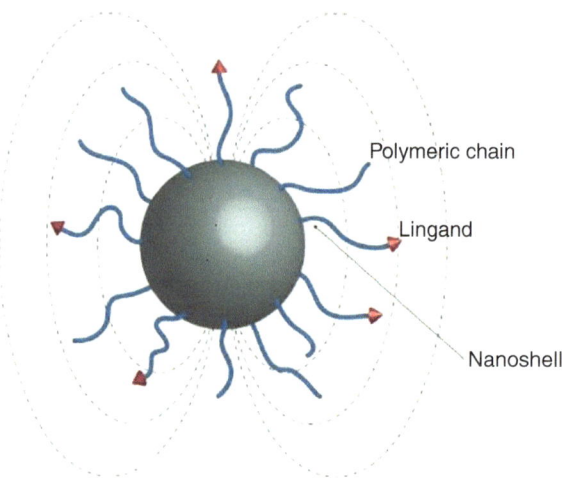

Fig. 6.1 Schematic representation of a polymeric coating nanosystem with magnetic properties

and targeting factors in a core-shell nanoparticles), just like iron oxide. There are of course other magnetic nanoparticles (elementals or compound alloys) more toxic than iron oxides. Examples of magnetic nanoparticles are Co, Ni, Cu, Ag, Fe, $MnFe_2$, and some core-shell nanoparticles $FePt/Fe_3O_4$. Nanoparticles of lipid composition are very popular as they are extremely biocompatible and cheap, can be developed in many different ways and in various sizes, and have great surface ratios to volume. Superparamagnetic iron oxides are distinguished in large superparamagnetic iron oxides with a diameter of 50–100 nm and ultra-small superparamagnetic iron oxide nanoparticles (USPIO) with a median diameter of 50 nm. These oxides are metabolized to elemental iron and oxygen with hydrolytic enzymes and then are used in the organism for other functions. Of course, iron oxides show relatively small magnetic susceptibility in comparison, for example, to metal iron nanoparticles that are oxidized too easily. Lastly, we can mention that toxicity, for example, in silver nanosystems, is reversely proportional to their size and smaller nanoparticles have larger toxic effects from their larger ones [10,15].

6.3.1   Surface Coating: A Way to Reduce Nanosystem Toxicity

Surface coating is the most important process in nanosystem production for bio-medicine applications, i.e., the process of covering-coating the nanoparticle nuclei with a biocompatible material, natural (like proteins) or synthetic (like dextran or polyethylene glycol). At this point, we should understand that the material chosen as coating material can be biodegradable, but this does not mean that the produced nanoparticle will be biodegradable too. The compounds used as surface coating are placed around the nanoparticle nuclei and create an extra layer that makes the nanoparticle invisible from the immune system.

General, nanosystem surface coating materials offer the following advantages:

- Colloidal stability under normal circumstances
- Functionalization enhancement
- Opsonization avoidance process (there will be an extensive reference in the following chapter)
- Resistance in phagocytosis from macrophages cells
- Mechanical stability
- Prolonged half-life
- Biodegradability
- Biocompatibility
- Oxidation resistance enhancement, regarding magnetic nanoparticles

Therapeutically speaking, functionality possibility is crucial when in need of a multifunctional particle (simultaneous targeting, selectivity, and treatment). Multifunctionality is related to bio-nanomaterials used as coating materials with the ability to connect chemically or in an electrostatic way with different molecules at the same time, so each one of these molecules will act autonomously and independently from the rest for the function desired (e.g., simultaneous cancer cell detection through MRI signal enhancement or through fluorescent in visible and chemotherapeutic cytotoxic factor targeted release in controlled release conditions). Polymer and synthetic chemistry development, during the past years, allowed the use of many compounds as surface coating materials.

Gold is a very popular coating since its chemistry is familiar and offers remarkable resistance to corruption and oxidation. Its biocompatibility is not desirable since gold does not exist in normal circumstances into the organism; therefore, it is considered as a foreign body and can cause immune reaction and toxic actions. Polysaccharides, i.e., large hydrocarbon molecules of chained or branched form (composed of many polysaccharide units linked together with condensation reactions), are extremely biocompatible and offer great possibilities to avoid phagocytosis, but they are structurally unstable reactions and can be dissolved in acid environment. Examples for polysaccharide coating are dextrans (composed of many glucose molecules).

Another type of surface coating that will be mentioned later on as multifunctional nanosystem coating is silicon-based coating. These surface coatings are used to protect nanosystems from lysosomal enzyme digestion, improving at the same time their mechanical properties and physicochemical stability. As these coatings create porous surfaces, their inner components might dissolve or be oxidized by interacting with nanoparticle components. Lastly, very useful materials for surface coating of metal or ceramic nanosystems or nanosystem adherence on their surface are polymers. The polymer that functions more often as a polymer material is polyethylene glycol (PEG). Other coating types are amphiphilic molecules (with hydrophilic heads and hydrophobic tails), like carboxylic acids and phospholipids, sol-gel-based coatings, etc. The Guidance on the Determination of Potential Health Effects of Nanomaterials Used in Medical Devices was published in 2015 by the Scientific Committee on Emerging and Newly Identified Health Risks (SCENIHR) which is authorized as the opinion of this committee. The risk assessment and aspects regarding the safety of medical devices that are composed of nanoparticulates are presented in this opinion of the SCENIHR committee.

The definition of what a medical device is was given by the directive 93/42/EEC as amended by Directive 2007/47/EC "any instrument, apparatus, appliance, software, material or other article, whether used alone or in combination, including the software intended by its manufacturer to be used specifically for diagnostic and/or therapeutic purposes, and necessary for its proper application, intended by the manufacturer to be used for human beings for the purpose of: – diagnosis, prevention, monitoring, treatment or alleviation of disease, – diagnosis, monitoring, treatment, alleviation of or compensation for an injury or handicap, – investigation, replacement or modification of the anatomy or of a physiological process, – control of conception, and which does not achieve its principal intended action in or on the human body by pharmacological, immunological or metabolic means, but which may be assisted in its function by such means." A new medical device regulation amended the above (2014) under negotiation (EC 2012), but the amendments that were proposed did not affect the already existing definition.

They are referred in the guidance report by the SCENIHR as nanomaterials that are used in medical devices as the following: carbon nanotubes, nanopaste hydroxyapatite powder, polymeric material, polycrystalline nanoceramics, and nanosilver or other nanomaterials used as coatings on implants and catheters and as an antibacterial agent [17]. The physicochemical characterization of the nanomaterials that the medical devices are composed of is included in ISO 10993-18: 2005 (*ISO 10993-17:2002. Biological evaluation of medical devices – Part 17: Establishment of allowable limits for leachable substances. ISO, Geneva, Switzerland*) and ISO 10993-19: 2006 (*ISO 10993-19:2006. Biological evaluation of medical devices – Part 19: Physicochemical, morphological and topographical characterization of materials. ISO, Geneva, Switzerland*). Concerning the toxicological evaluation of nanomaterials, the ISO/TR 13014:2012 describes the relevant guidance. We refer below some of the crucial parameters that should be taken into consideration during the evaluation process of nanomaterials used for medical devices: the size and size distribution, chemical composition, particle and mass concentration, charge (ζ-potential), surface properties, pH, viscosity, etc. as described in the relevant document. The methods of their characterization have been already presented in the Chaps. 2 and 4. The electron microscopy methods are proposed as more appropriate to characterize nanoparticles as in the SCENIHR document referred.

6.3.2   Clearance Time: The Most Critical Parameter of Toxicokinetics

Toxicokinetics provide information regarding the behavior of nanoparticulates when they arrive in specific organs and tissues. Their ADME (absorption, distribution, metabolism, and excretion) profile of nanoparticles is directly related to their toxicokinetic profile. The metabolism of particular nanomaterials has been published. It is considered that the kidney is an important organ among others, because of the excretion of nanomaterials, while the spleen and liver are also parts of their total toxicokinetic profile. Some nanoparticles showed rapid uptake in the kidney in comparison with the liver when administrated intratracheally, for example, Au nanoparticles with a size of 1,4 nm, while after intravenous administration, the liver is found to be the accommodation place of these nanoparticles. However, the route of administration seems to be essential for their accumulation in organs and consequently affects their toxicokinetic profile [13]. EFSA [2] underlines the importance for evaluating nanoparticles in specific organs like the liver, kidney, and spleen because of their high uptake from this organ.

A basic parameter for nanosystem viability into the human organism is the maximum time they can preserve their properties under normal in vivo conditions. This is the clearance time, i.e., the time until full nanosystem clearance from the organism.

From the time nanoparticles enter the bloodstream/system circulation, most possibly penetrating some barriers (often by intravenous infusion), are very soon covered from systemic circulation protein components, like plasma and complementary proteins, with a process called opsonization. Opsonization makes nanoparticles detectable from the main human defense mechanism, MPS, and then they are submitted for phagocytosis. An important part of the nanosystems has been cleared from the circulation system in less than 15 min. Of course the nanosystem clearance rate is a function of various parameters related to their structure and functions, like:

- Nanoparticle size
- Shape
- Nanosystem surface charge
- Surface hydrophobicity
- Surface functional group number and kind
- Presence of surface coating materials

For example, small nanoparticles can get away easier from the endothelial network system, and for this reason, they have greater clearance time and extended action and, therefore, more unwanted and possible toxic actions. Therefore, these anionic particles have negatively charged surface and are similar to the cell membrane; they are absorbed and biodistributed very easily during endocytosis process. Cationic particles (i.e., cationic liposomes) can connect electrostatically with DNA molecules for gene transport and great toxicity that depends on the size of their surface charge. Generally, the more cationic the nanoparticle, the more toxic it is for the organism. The hydrophilic surfaces, i.e., the hydrophilic coating materials like dextran and polyethylene glycol, offer a potential hydrophilic and neutral chain cloud in the nanoparticle surface that repels plasma proteins and prevent quick clearance from the systemic circulation. Also, nanofibers can penetrate smooth tissues, like the nervous system, blood vessels, and lungs, and create topical inflammatory reaction. Lung reaction resembles the professional amianthus disease that was due to amianthus fibers and caused acute pulmonary reaction and chronic pulmonary fibrosis. Therefore, the clinical doctor should be suspicious with patients presenting fever with pulmonary inflammation that is not treated with antibiotics and develops chronic disease if the patient has a history of contacting nanotechnology materials [10,15].

Other diseases mentioned are vascular diseases, for example, arteritis in the peripheral blood vessel or bowel arteritis and hepatitis. Parkinson's and Alzheimer's diseases and some other neurodegenerative diseases or brain vasculitis could be due to nanotoxicity, and the clinical appearance is representative to the infected section. Arthritis that resembles rheumatic diseases, inflammatory bowel diseases, and bowel cancer can be related to this disease, and skin diseases are related to HIV and Kaposi sarcoma. It can be said that nanotoxicity belongs to the diseases related to the environment and human way of life. The experience is limited, and the small fiber behavior resembles the behavior of known chemical fibers, like amianthus fiber, and of forms called prions and affects human neuron and muscle systems, causing the known spongiform encephalopathy. The World Health Organization reports nanotoxicity as an environmental disease. It is possible that in the next years, it will be in most medical specialties and all care levels. Figure 6.2 presents the

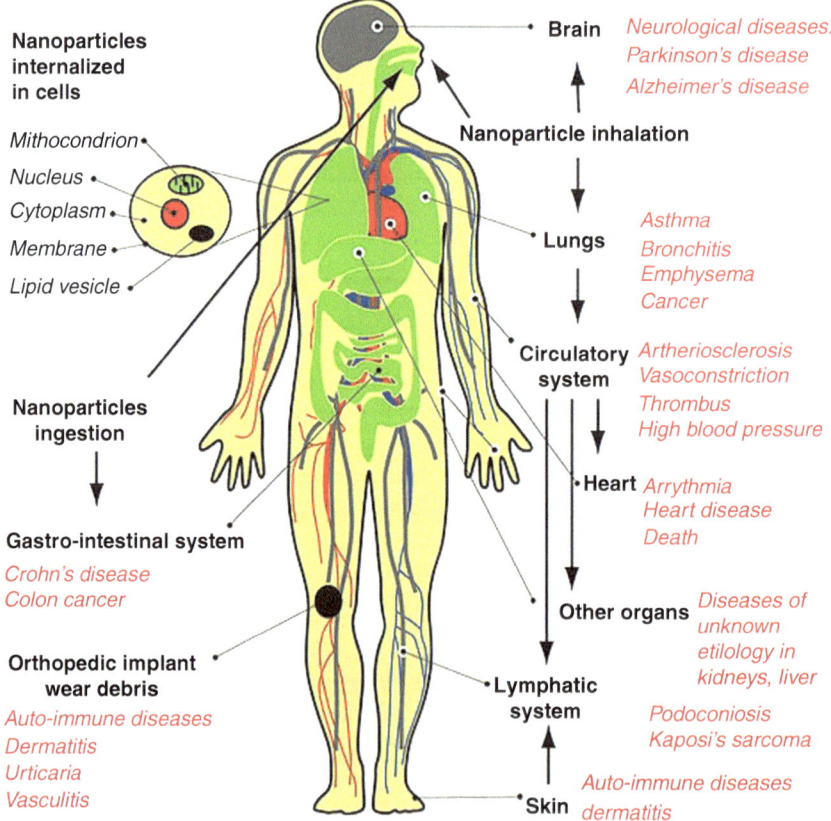

Diseases associated to nanoparticle exposure

C. Buzea, I. Pacheco, & K.Robbie, Nanomaterials and nanoparticles: Sources and toxicity, Biointerphase 2 (2007) MR17-MR71

Fig. 6.2 Diseases associated with human exposure to nanoparticles (Adapted from Buzea et al. [1] with permission from Springer)

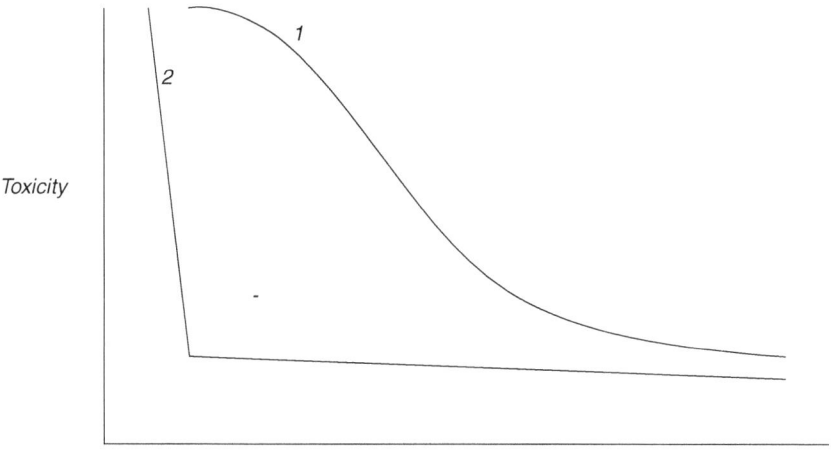

Size of nanoparticles

Fig. 6.3 Relationship between toxicity and particle size. *1.* Sigmoid curve. Progressive increase of toxicity by reducing the particle size. *2.* Below a certain particle size, small variation in particle size significantly enhances toxicity

diseases related to nanotoxicity. Nanotechnology products currently in the market are hundreds and relate to everyday life. Mostly they consisted of carbon nanotubes and titanium oxide nanoparticles, but until today, no detailed epidemic study of the short- and long-term use in the human organism was carried out.

We must mention that the main reasons of toxicity appearing in materials of nanoscale dimensions are their structure. Figure 6.3 presents the relationship between toxicity and particle size. We can observe in the sigmoid curve (1) the progressive increase of toxicity in relation to particle size, while beneath a specific particle size, see curve (2), a small particle size change induces an important toxicity increase.

Summarizing, it is obvious that the clearance process of nanoparticles from the circulation has occurred via the spleen and the liver [7,8].

6.4   Summary

The toxicity of nanoparticulate systems seems to be a result of their great total surface area and of their dimension. It should be taken into consideration that the nanotechnological products might be developed without the manufacturers strictly following the necessary rules for their production, transport, and storage.

The physicochemical properties of nanomaterials and nanosystems like size, size distribution, charge, shape, and surface property stability might be involved into their toxicity.

The nanotoxicity in molecular level is defined as biotoxicity.

The route of administration seems to be essential for the accumulation of nanoparticles in organs and consequently affects their toxicokinetic profile.

An important aspect that should be taken into consideration is the clearance rate of nanoparticles from the organism. The clearance of nanoparticles from the blood has occurred mainly via the spleen and the liver.

References

1. Buzea C, Pacheco II, Robbie K (2007) Nanomaterials and nanoparticles: sources and toxicity. Biointerfaces 2:MR17–MR71
2. EFSA (2011) Guidance on the risk assessment of the application of nanoscience and nanotechnologies in the food and feed chain. EFSA J 9:2140
3. Ehmann F, Sakai-Kato K, Duncan R et al (2013) Next generation nanomedicines and nanosimilars: EU regulators' initiatives relating to the development and evaluation of nanomedicines. Nanomedicine (Lond) 8:849–856
4. Kewal KJ (2008) The handbook of nanomedicine. Humana Press, Basel
5. Kreuter J (2001) Nanoparticulate systems for brain delivery of drugs. Adv Drug Deliv Rev 47:65–81
6. Landsiedel R, Kapp MD, Schulz M et al (2009) Genotoxity investigations on nanomaterials: methods, preparation and characterization of test material, potential artifacts and limitations – many questions, some answers. Mutat Res 68:241–258
7. Lankveld DPK, Rayavarapu RG, Krystek P et al (2011) Blood clearance and tissue distribution of pegylated and non-pegylated gold nanorods after intravenous administration in rats. Nanomedicine 6:339–349
8. Lipka J, Semmler-Behnke M, Sperling RA (2010) Biodistribution of PEG-modified gold nanoparticles following intratracheal instillation and intravenous injection. Biomaterials 31:6574–6581
9. Mehta D, Bhattacharya J, Matthay MA et al (2004) Integrated control of lung fluid balance. Am J Physiol Lung Cell Mol Physiol 287:L1081–L1090
10. Muller J, Huaux F, Moreau N (2005) Respiratory toxicity of multi-all carbon nanotubes. Toxicol Appl Pharmacvol 207:221–231
11. National Nanotechnology Initiative (NNI) (2006) NNI: environmental, health and safety research needs for engineered nanoscale materials. NNI White paper issued by the Office of the President of the United States, September, 2006
12. Nel A, Xia T, Madler L et al (2006) Toxic potential of materials at the nanolevel. Science 311:622–627
13. Oberdörster G (2010) Safety assessment for nanotechnology and nanomedicine: concepts of nanotoxicology. J Intern Med 167:89–105
14. Ryman-Rasmussen JP, Riviere JE et al (2006) Penetration of intact skin by quantum dots with diverse physicochemical properties. Tox Sci 91:159–165
15. Sato Y, Yakoyama A, Shibata K et al (2005) Influence of length on cytotoxicity of multi-walled carbon nanotubes against human acute monocytic leukemia cell line THP-1 in vitro and subcutaneous tissue of rats in vivo. Mol Biosyst 1(2):176–182
16. Tratnjek PG, Jonhnson RL (2006) Nanotechnologies for environmental cleanup. Nanotoday 1:44–48
17. Wijnhoven SWP, Peijnenburg WJGM, Herberts CA et al (2009) Nano-silver – a review of available data and knowledge gaps in human and environmental risk assessment. Nanotoxicology 3(2):109–138

Chapter 7
Regulatory Framework for Nanomedicines

Abstract The nanotechnological products that are released in market are carefully evaluated in the United States by the National Agency for Food and Drugs that is called the Food and Drug Administration (FDA), while in Europe by the European Medicines Agency (EMA). These organizations act according to proposals and recommendations of scientific committees that suggest changes and modifications of the regulatory framework according to scientific, technological, and social facts keeping into consideration as the main concern, the public health. These organizations develop the regulations for law applications related to medicines and medicinal products. The "copies" of nanotechnological medicines that are candidates to release in market are called as nanosimilars. They are *off-patent* nanotechnological products, and their similarity is considered as an emerging issue to be discussed within the scientific community and moreover in the regulatory bodies which are responsible for their approval. An important issue from the regulatory point of view is *innovative excipients* that include the self-assembled nanostructures such as liposome, micelles, dendrimers, etc. incorporating bioactive molecules.

Keywords Regulatory framework • Nanotechnological products • Food and Drug Administration (FDA) • European Medicines Agency (EMA) • Nanosimilars

7.1 Regulatory Framework for Nanotechnological and Nanobiotechnological Medicine Approval

Laws are the society's basic structural characteristics and contribute to its development. Laws are developed from the national parliament in each country, and their applications are controlled by regulations. In the United States, the National Agency for Food and Drugs that develops the regulations for law applications related to health, drugs, and food is called the Food and Drug Administration (FDA), while in Europe, the organization that develops the regulatory framework for pharmaceutical product approval is the European Medicines Agency (EMA). These organizations act according to proposals and recommendations of various scientific committees that suggest relative changes and modifications of the regulatory framework

© Springer Science+Business Media Singapore 2016   189
C. Demetzos, *Pharmaceutical Nanotechnology*,
DOI 10.1007/978-981-10-0791-0_7

according to scientific, technological, and social facts and the pharmaceutical product or formulation effect in human health, i.e., safety and efficacy.

The legislative framework for nanotechnological and nanobiotechnological medicinal product approval is constantly changing and developing [14]. High-technology nanotechnological product developments even for everyday applications create the need for a legislative framework that will be updated with technological evolutions [2,10]. Also, we must mention that social groups' reflexes in the field of nanotechnology product safety use are still very weak. It seems that public awareness is in primitive stage and limited mostly into scientific reports and papers that common people cannot understand easily. The present chapter describes the necessary conditions for nanotechnology health product approval. The commercial exploitation of nanotechnological products in the United States is faster in comparison to the EU. This is mainly due to the investment status established in the United States in relation to innovative products and idea promotion, in comparison to the introvert character of the EU countries and Europe in general. Generally there is a need of conductive conversations and good natured effort for the evolution of the present legislative framework for approving new and innovative pharmaceutical products and diagnostics, especially those developed based on nanotechnology [14].

The legislative framework that is developed through regulations is directly related to the scientific approach of nanoparticle quality and safety but mainly with challenges in various product development and production stages. The new nano-product production abilities concern nanotechnology application to both hydrophobic and hydrophilic bioactive molecules in various solubility degrees in biological fluids [6]. Nanoparticles are developed in various techniques and methods that are evaluated from the approval mechanisms, before final product authorization for market placing.

Crystalline bioactive molecule nanoparticles present advantages in their solubility and dissolution rate. According to the Kelvin equation, particle nanodimension leads to greater solubility:

$$RTln\left(\frac{S}{S_0}\right) = \frac{2\gamma V_m}{r} \tag{7.1}$$

where R is the gas molecular constant, T the absolute temperature, S the solubility of spherical particles with radius r, S_0 the large crystal particle relative solubility, γ the interfacial tension, and V_m the partial molar volume.

Nanoparticle parameters and characterization methods required for approval process for safety and effective final product placement in the market are presented in Table 7.1.

The parameters and physicochemical methods for bioactive molecule nanoparticles and nanosystem characterization are required and controlled for accuracy and repeatability from the approval mechanisms prior to product authorization process.

Table 7.1 Parameters and characterization methods for the approval of nanoparticles

Parameter	Method
Size distribution of and shape of nanoparticles	Dynamic light scattering, photon correlation spectroscopy, electron microscopy
Surface area, porosity	Gas absorption
Surface charge, hydrophilicity	Electrophoresis, ζ-potential
Surface properties	Static secondary ion mass spectroscopy
Analysis of surface elements	X-ray photoelectron spectroscopy
Density	Pycnometer/densitometer
Molecular weight	Size exclusion chromatography
Purity	Spectroscopy
Polymorphism, crystalline state	X-ray diffraction, thermal analysis
Solvent residues	Gas chromatography
Drug release	Drug release studies, turbidity

7.1.1   The Regulatory Framework for the Nanobiotechnological Products in the United States and the Role of the FDA

Nanotechnology product supervision is applied in the United States from the FDA that exercises the legislative supervision by the Office of Science and Health Coordination through the Office of Commissioners [16]. The work groups suggest scientific proposals and regulations for study, development, and promotion of new and innovative pharmaceutical products. Examples of medicines approved from the FDA are products with nanocrystals, liposomal products, albumin products connected with nanoparticles for anticancer drug production, pharmaceutical products with microemulsions and micelle technology, and pharmaceutical products with polymer production (Table 7.2).

An important parameter for product safety is the nanoparticle size and size distribution measurement as well as the potential determination [12]. These studies are related to their toxicity as large nanoparticles when forming aggregations or/and agglomerations may develop thromboembolic episodes, crucial to life. The Center for Drug Evaluation Research also known as CDER requires the same controls regarding new medicine safety to be applied to nanotechnological products. According to the Chairman of research policy and conformation of CDER pharmaceutical science office, appropriate regulations should be taken under consideration for new nanotechnological pharmaceutical products in order to be safe and effective. We can mention as important decision centers inside the FDA the following: Center for Biological Evaluations and Research (CBER), Center for Veterinary (CVM), and Center for Device and Radiological Health (CDRH). It should also be mentioned that the office for control and compliance according to the applicable regulatory framework is the Office of Regulatory Affairs (ORA).

Table 7.2 Examples of medicines based on nanoparticulate technology approved by the FDA

Brand name	Active ingredient	Indications	Formulation	Company
Microemulsion and micelle products				
Oraqix	Lidocaine and prilocaine	Local dental anesthetic	Gel, 2.5%/2.5%	Dentsply pharmaceutical
Lipofen	Fenofibrate	Lipid regulator (antilipidemic)	Capsules	Cipher pharmaceuticals
Gengraf	Cyclosporin	Protection of rejecting organ transplants	Capsules	AbbVie pharmaceuticals
Polymeric products				
Oncaspar	Pegaspargase	Acute lymphoblastic leukemia	Injectable	Sigma Tau PharmaSource Inc.
Sylatron	PEG – a-interferon 2b	Chronic hepatitis C	Injectable	Schering-Plough corporation
PEGasys	PEG – a-interferon 2b	Chronic hepatitis C	Injectable	Hoffmann-la Roche
Somavert	Pegvisomant	Acromegaly	Injectable	Pharmacia and Upjohn

Nanopharmaceutical product approval processes follow nowadays the already existed certification procedure, since there is no special regulatory framework for nanotechnological pharmaceutical products. Some important issues should be taken into consideration before the approval of a nanobiotechnological product, such as:

- The most appropriate techniques that should be applied for accurate physico-chemical characterization.
- The stability studies of nanopharmaceutical product in short- and long-term period time.
- The analytical methods for measuring the pharmacokinetic profile of the products should be validated and accurate.
- The scale-up process should be determined in details in order to avoid unexpected phenomena.
- The quality of the raw ingredients that have to be used for producing the final product.
- The adequate techniques that efficiently determine the kinetics and the release profile of the bioactive molecule from the nanocarrier.

While nanotechnology and its application importance will be more noticeable in the future, FDA has already approved products based on nanotechnology, like imaging factors and nanosystems in sun block filters. FDA, according to complex controls and procedures, has approved nanopharmaceutical products (Table 7.2), like liposomal products, tissue imaging products (including gadolinium) used in the magnetic resonance imaging (MRI), iron oxide products, and contrast agents.

There are nanopharmaceutical systems like liposome or lipidic systems that are considered to be important for bioactive molecules like doxorubicin, daunorubicin, amphotericin B, cytosine arabinoside, etc. that are incorporated into lipidic nanocarriers. In some other products like sirolimus, NanoCrystal technology was applied and received an FDA approval, while fenofibrate, aprepitant, etc. are some of the pharmaceutical product examples whose formulation has changed by using NanoCrystal technology, also approved by the FDA. Another product is Vivagel® that is a dendrimer and binds to HIVgp 120 proteins and now in clinical trials.

7.1.2   The Regulatory Framework for the Nanobiotechnological Products in the EU and the Role of the EMA

As previously mentioned in the European Union, the European Medicines Agency (EMA) is responsible for the regulatory provisions concerning the national regulatory provisions for biotechnological and nanotechnological products. In the Agency there are six committees that are presented in the following:

- Committee for Medicinal Products for Human Use (CHMP)
- Committee for Medicinal Products for Veterinary Use (CVMP)
- Committee for Orphan Medicinal Products (COMP)
- Committee on Herbal Medicinal Products (HMPC)
- Pediatric Committee (PDCO)
- Committee for Advanced Therapies (CAT)

The committees mentioned above meet in a monthly basis and are composed of scientists assigned by the member countries. The therapeutic and pharmaceutical product evaluations that are currently in the market or are about to be available in the European Union are based on the scientific developments and social needs according to safety and efficacy regulations and European legislation and especially the guideline 2001/83/EC.

These Committees based on the above criteria and European directions decide – considering the risk/benefit ratio – for a therapeutic product according to relative studies.

The therapeutic product development process demands the production compliance to guidelines and includes [4,16]:

- Preclinical studies in animals to study the toxicity
- Clinical studies of various levels in patients and healthy volunteers to study the effectiveness and to define the appropriate dose
- Post-approval studies concerning the pharmaceutical or therapeutic product safety and efficacy monitoring when already in the market

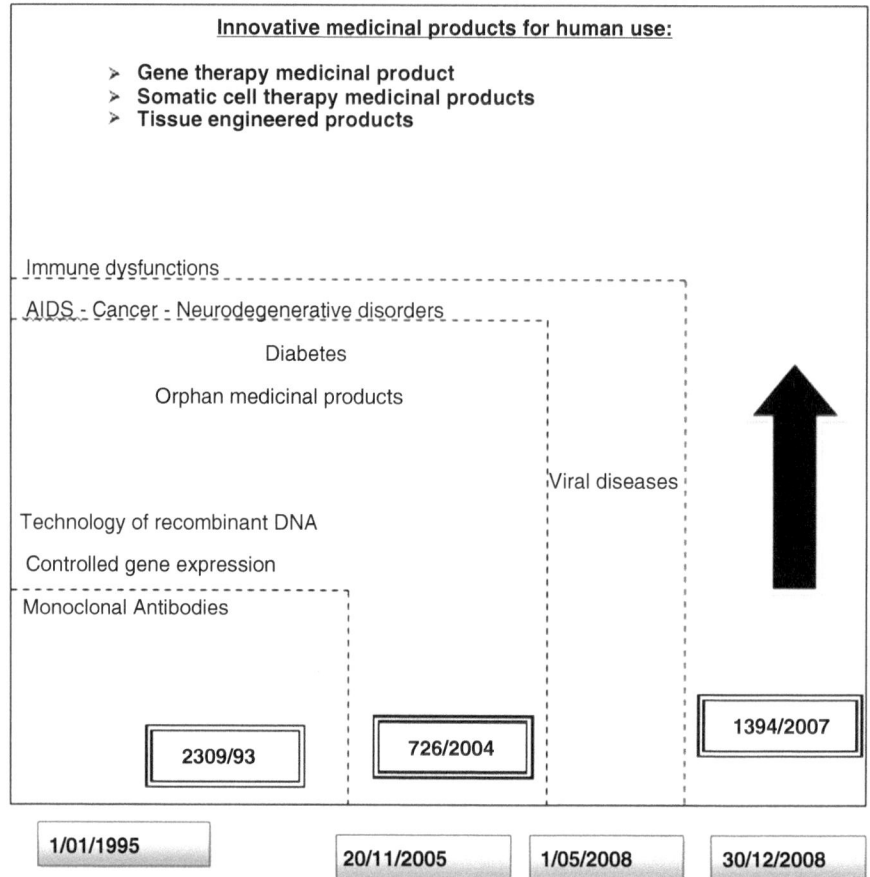

Fig. 7.1 Regulatory extension of the centralized procedure for innovative medicinal products for human use

We should mention that the therapeutic product is something different than the classic medicine. Therapeutic products belong to a class of biotechnological medicinal products that do not include bioactive molecules but biological molecules such as proteins, peptides, monoclones, large molecules, or even living cells whose physicochemical behavior and especially the structural characteristic behavior (depending on the environment and immunogenicity) are extremely important factors for safety and efficacy [2, 5, 8, 10].

We should also mention that pharmaceutical and therapeutic product post-approval studies are extremely important and relate to the science of pharmacovigilance and safe use of such products. The European legislation according to the regulation EU No 1235/2010 of the European Council and the European Committee modified in December 2010 mentions pharmacovigilance for pharmaceutical products for human use. The European Committee (EC) regulatory provisions (EC) No

2309/1993, (EC) No 726/2004, and (EC) No 1394/2007 (Fig. 7.1) for advanced therapies mention that pharmacovigilance is necessary for the protection of public health in order to identify and evaluate the medicine and therapeutic product side effects for human use [9].

The post-approval monitoring of pharmaceutical and therapeutic product safety is necessary for the continuation –or not – of the product and the identification of possible side effects that were not detected during clinical trials and are not mentioned in the patients' leaflet.

New therapy development with products characterized as advanced therapy medicinal products (ATMP) in the European environment is recognized from the scientific community and from international approval agencies as important. Safety and efficacy of such products, like gene therapies, body cells, therapeutics through tissue mechanics, etc., demand regulatory provisions and probably regulatory framework that will take under consideration the new innovative scientific and technological evolutions. The review of approval procedures is required to be according to the scientific evolutions and the existing regulatory framework.

The EMA committee, referred as CAT, proposes recommendations and settings using the central approval procedure for advanced therapeutic products, the Marketing Authorization Applications (MAAs), with the positive outcome of the Committee for Medicinal Products for Human Use (CHMPs), publishing the appropriate regulatory framework. It is also important to mention that the strategy for the regulation application for innovative therapeutic products includes nanotechnology products produced for therapeutic purposes – there will be an extensive report in the following.

Therapies like gene therapy medicinal products (GTMPs) and tissue-engineered products (TEPs) should be subjected to regulations, not only during production but when in the market as well.

In our opinion and from all the above, a lot of work is required for regulatory provisions related to nanotechnology products and innovative therapies, since the scientific and technological orientations in the field of nanotechnology take advantage of the biomaterial special properties in mesoscopic level and matter quantum mechanics properties [15].

The approval process follows the processes required for new medicines (mutual, national, central approval process for pharmaceutical products), having developed the appropriate committee for safety and efficacy control for new innovative medicines that as we mentioned is called the Committee for Advanced Therapies (CAT). A review of the existing and an estimation for the future environment in the European Union regarding the nanotechnological product approval regulatory framework as well as comparative data regarding the European Union and the United States in the field of nanotechnological products are in the reference section.

The regulatory provisions seem to be a very interesting matter for the body of experts that define the approval regulatory framework. This is because the pharmaceutical products in the nanodimensions that are produced present different properties and behavior than the one of the materials in classic dimensions. The efforts for mapping the regulatory framework regarding the nanotechnological pharmaceutical products are very important. FDA and 22 federal approval agencies developed the National Nanotechnology Initiative (NNI) in order to coordinate the actions for approving nanotechnological pharmaceutical and therapeutic products. Another important aspect for the approval mechanisms' effective role is the terminology determination related to nanotechnology.

The term nanotechnology should be defined exactly with no variations in the size limits and parameter value dispersion but also in method evaluation for pharmaceutical and therapeutic substance nanoparticles and nanocarrier physicochemical characteristics.

Based on this idea that is considered as definitive, the precise subject determination where the regulations and the future legislative framework should be referred to, different perspectives were stated from the American Standards for Testing and Materials (ASTM), British Royal Society, and Royal Academy of Engineering, and the US Congress and even the National Strategy for Combating Terrorism (NCST) gave different definitions on the nanotechnology subject and effect size for protective regulatory framework against nanotechnology product terrorist actions. We should mention that the differences on nanotechnology term according to nanoparticle size lead the manufacturers to therapeutic nanotechnology and in the future nanobiotechnology product attentive production based on the specific regulations.

The work of Wagner and his colleagues [16] in the journal of *Nature Biotechnology* [16], entitled *The emerging nanomedicine landscape*, mentions the nanobiotechnology applications and the need to develop new regulatory provisions. More in particular they mention that "…there is a continuous consultation between scientists, industry and experts that participate into approval mechanisms regulatory committees regarding the need for new regulations for the therapeutic products and nanobiotechnology/nanotechnology drugs pharmacokinetic properties [16]."

In the European Union, for the products mentioned above, there is not a clear regulatory framework and the guidelines are proposed in the article 1.2(b) Directive 2001/83/EC of the European Parliament and of the Council of November 6, 2001, on the Community code relating to medicinal products for human use. Off. J. Eur. Union, November 28, 2001, as amended, consolidated version: October 5, 2009.

The different approaches for the term nanotechnology may push to release to an agreement between the United States, the European Union, and Japan that regarded the pharmaceutical nanotechnology as a rising scientific field that should be defined in order to develop the relative regulatory framework. The approval framework development and application are related to science and technology evolutions and have a powerful and continuous presence. The final product safety and efficacy upgrade will contribute to new tool development that will target toward the nanoparticles, and of course the final nanotechnological products, physical and chemical properties, emphasizing in parameters like nanoparticle size and size distribution and surface characteristics and developing intersurfacial phenomena, in combination with other analytical approaches regarding excipients, impurities, and organic residues. Also, the approval mechanism guidelines for nanoparticle toxicity in preclinical studies and the guidelines Q8 and Q9 of the International Conference for

Harmonization (ICH) are characterized as very important in the development of nanopharmaceutical products for therapeutic purposes.

Lastly, in our opinion, what seems to be an absolute prerequisite for new nano-pharmaceutical product approval is public partnership and sensitive social groups' participation to accept these innovative therapeutic, diagnostic, and imaging products. The health systems should take under consideration these specific social groups to ensure the public's approval for nanopharmaceutical product development and their implementation to health systems.

7.1.2.1 Commission for Advanced Therapies (CAT)

The Commission for Advanced Therapies (CAT) was founded recently (2007) and was composed according to the regulation (EC) No 1394/2007 and proves that the European Agency for Pharmaceutical Products considers the evolutions in science, especially in biotechnology, nanotechnology, regenerative and molecular medicine, as well in tissue mechanics [9].

More in particular, the committee for advanced therapies is an interdisciplinary scientific committee that collected some of the finest experts available in the European area to evaluate the innovative therapy quality, safety, and efficacy and coordinates with the scientific evolutions in this field. The new scientific approaches in cellular and molecular biotechnology have led to advanced therapy development. The rapidly growing field of biomedicine offers new opportunities for disease treatment and human body congenital malfunctions. Advanced therapy products should have properties for disease treatment or prevention in order to be used or administered in humans for replacement, improvement, or modification of normal functions. They should have pharmacological, immunological, and metabolic action and be classified as biological drugs according to the Appendix 1 of guideline 201/83 EC of the European Parliament and the Council of November 6, 2001, on the Community codes relating to medicinal products for human use, in combination to pharmaceutical product definition in the article 1(2).

7.1.2.2 Innovative Pharmaceutical Products

Innovative pharmaceutical products differ from pharmaceutical products available in pharmacies and are open to the public and hospital pharmacies. According to the guideline 201/83/EC and regulation (EC) No 726/2004, advanced therapy medicinal products (ATMPs) are any of the following pharmaceutical products for human use:

- Gene therapy medicinal products
- Somatic cell therapy medicinal product
- Tissue-engineered products

More in particular, according to the European Community (EC) guideline No 668/2009 emphasizing the terms for advanced pharmaceutical products, a gene therapy medicinal product is a biological pharmaceutical product that has the following characteristics:

- Includes the therapeutic agent that is composed from recombinant nucleic acid that is used or can be administered in humans in order to regulate, fix, replace, add, or erase a genetic sequence.
- The therapeutic, prophylactic, or diagnostic factor is directly related to the recombinant nucleic sequence included or to the product of this genetic sequence expression.

According to all the above and the European Community guideline mentioned before, gene therapy medicinal products do not include vaccines against infectious diseases that are subjected to the Committee for pharmaceutical product for human use regulatory principle.

Somatic cell therapy medicinal product is a biological pharmaceutical product that has the following characteristics:

- Includes or is composed of cells or tissues that have been modified so that their biological characteristics, their physiological/normal functions, or their structural properties that are important for their clinical use have been modified either from cells or tissues that are not destined to be used for the same basic function(s) of the recipient and the donor [2,8,10].
- Have properties that target disease therapy, prevention, or diagnosis due to pharmacological, immunological, or metabolic action in cells or tissues [2,8,10].

According to scientist conversations participated from all European Union countries in April 3, 2009, meeting the view that was formed, the pharmaceutical product classification that harmonizes with the committee's regulations is very important since drugs are not inactive materials, but cells and bionanomaterials that respond to external stimuli. The centralized procedure obligatory provision extension on innovative pharmaceutical product regulations according to European Medicines Agency articles and guidelines/regulations is presented in Fig. 7.1.

Furthermore, according to the regulation (EC) No 1394/2007, the central procedure was obligatory for these innovative products. Also, the following were decided:

- Knowledge and information grouping regarding specific products with community experts' background knowledge
- Requirement harmonization and post-evaluation
- Access assurance for all consumers to these products in the market
- The existent regulation principles for drugs applied in advanced therapies, i.e., quality, assurance and effectiveness, authorization, and post-approval vigilance (pharmacovigilance)

7.1.2.3 Approval Process for Innovative Pharmaceutical Products

The estimations for an innovative pharmaceutical product approval are based on scientific criteria that define if these medicines fulfill or not the necessary requirements for quality, safety, and efficacy (according to European Union regulations and especially according to the guideline 2001/83/EC). These processes assure that the innovative medicine risk benefit ratio is in consumers' favor. For the innovative pharmaceutical product development process, the following general steps should be followed according to the strategies for pharmaceutical product regulation:

• Preclinical development phase
• Clinical development phase
• Post-approval development phase

It should be noted that the last post-approval phase is of great importance, since it is related to public health and is defined as pharmacovigilance.

Apart from all the above that also apply with nuances for the rest of the pharmaceutical products for human use, the following requirements according to the European Committee 2009/120/AC on September 14, 2009, for the European Council 2001/83/EC guideline modification, regarding the Community code establishment for pharmaceutical products for human use, should be applied. For the advanced therapy pharmaceutical products, the following apply:

1. Information for all the raw ingredients used for the bioactive molecule production should be provided. This includes the products for human or animal cell genetic modification, and when they are applicable for their further culture and modification, taking under consideration the possible absence of cleansing.
2. For the products that include a microorganism or virus, the following data should be mentioned and described: genetic modification, sequence analysis, virulence attenuation, tropism for specific tissues and cell types, microorganism or virus cycle dependence, gene strain pathogenesis, and characteristics.
3. Processes related to infectious parameters and to the product should be described in the relative section of the authentication, especially if the virus is capable to be transcribed or the carrier is designed to be transcribed-free.
4. Quantification of plasmids' various forms are required and they should be taken under consideration through the product's life span.
5. For the genetically required cells, the cell characteristics prior and post modification are required. Apart from the special requirements for gene therapy pharmaceutical products, the quality requirements for somatic cell therapy medicinal products and tissue-engineered products should also be applied.

It seems that structure, function, and pharmacological properties of the advanced therapeutic properties based on nanotechnology, biotechnology, and molecular medicine and therefore their evaluation from the regulatory authorities responsible for quality, safety, and efficacy are questioned. For example, the Committee for Medicinal Products for Human Use in 2006 underlined the need for interdisciplinary approach in order to accept new development standards and to adjust the methodology for testing nanotechnological bioactive molecule advanced delivery systems. This flexibility in adopting the appropriate scientific tools to determine a complete regulatory framework is applied from the Committee for the European Medicines Agency innovative pharmaceutical products. On the other side, the scientific community responds in this challenge with numerous publications related to the management of nano- and biomaterials with astonishing physicochemical properties that reflect the improved biological actions to treat incurable and degenerative diseases.

7.2   Nanosimilars

Nanosimilars are considered as the "copies" of the *off-patent* nanotechnological therapeutic products. They are composed of the bioactive molecule and a nanocarrier which could be characterized as *innovative excipient*. It is important to point out as mentioned in Chapter 5 that the self-assembled supramolecular structures like liposomes, lipidic nanocarriers, polymerosomes, dendrimers, etc. could be defined as *innovative excipients*. In light of this approach, nanosimilars belong to the area between generics and biosimilars in terms of their evaluation process by the medicinal agencies. Moreover, the technological complexity of nanomedicines is considered as a barrier to develop identical "copies" of the prototype. However, the similarity between prototype and nanosimilar is a demand, and efforts should be focused on this approach in order to develop new and effective analytical tools for proving nanosimilarity. The approaches for proving similarity cannot be applied to nanosimilars as in the case of generic medicinal products (bioequivalence studies), while the structural complexity and the immunogenicity of biosimilars seem to be the major concern in the manufacturing process [1,7,11].

According to the European Medicines Agency (EMA), *a generic medicine is defined as a medicine that is developed, as the reference medicine and it contains the same active substance, and it is used at the same doses to treat the same diseases. Generic medicines are manufactured according to the same quality standards as all other medicines* (EMA/393905/2006 Rev. 2). Furthermore, the EMA defines that "a similar biological or "biosimilars" medicine is a biological medicine that is similar to another biological medicine that has already been authorized for use (EMA/837805/2011). Biological medicines are medicines that are made by or derived from a biological source, such as a bacterium or yeast (EMA/837805/2011). They can consist of relatively small molecules such as human insulin or erythropoietin, or complex molecules such as monoclonal antibodies." The Committee for Medicinal Products for Human Use (CHMP) has provided guidance through several guidelines, while advices requested by EMA are under discussion concerning the development and regulation approaches of biosimilars. EMA published an open consultation platform for biosimilar medicinal products on serious issues like the immunogenicity assessment of therapeutic proteins, the quality issues of biosimilar medicinal products containing proteins as active substance, and nonclinical and clinical issues of similar biological medicinal products containing monoclonal antibodies. The current regulation environment in EU regarding biosimilars has been extensively presented in a published work by Tsiftsoglou and coworkers [13]. Guidelines for nanosimilars in proportion/in analogy to biosimilars are posted by the EU in October 11, 2011. In an effort to contribute to the scientific discussion, Demetzos and Pippa [3] published a reflection note which deals with the approach to approve nanosimilars by the medicinal authorities proposing new analytical tools that should be carefully applied and evaluated.

The scientific discussion on *nanosimilarity* is an emerging approach that should be carefully preceded by the scientific community. Nanosimilarity is proposed as a new definition and projects the efforts that should be done in establishing the process of the evaluation, the similarity with the *off-patent* nanomedicine in terms of its physicochemical characteristics. It is important to point out that the surface coating and consequently the physicochemical profile of a nanosimilar as well as of a nanomedicine should be taken into consideration. However, EMA has published a reflection paper on general issues for consideration regarding the parental administration of nanomedicines with surface coating (EMA/325027/2013). In this reflection paper, EMA points out that the coating affects the pharmacokinetics and the biodistribution and the stability of the medicine, as well as the surface interactions with other biomolecules and the potential consequences of such interactions. According to the EMA/CHMP/SWP/100094/2011, "variation in mean/median size and size distribution and/or the accuracy of methods employed for nano-sizing may result in the generic product displaying different physicochemical properties leading to a different biopharmaceutical profile in respect of pharmacokinetics and biodistribution. This has the potential to significantly impact on safety/efficacy in comparison to the reference product" [3]. It is obvious that the physicochemical properties (i.e., size, size distribution, ζ-potential, etc.) of nanosystems are considered as crucial concerning their effectiveness. However, the full documentation profile in terms of the physicochemical characteristics provides evidences regarding the similarity of the nanosimilar product to that of the reference nanomedicine and is required for its approval. It should be noted that the minimum requirements of nanosimilar products are the highly similar physicochemical characteristics.

The ADME profile of the nanocarrier (i.e., liposomes, micelles, etc.) influences the pharmacokinetics of the bioactive molecule as the "Reflection paper on the data requirements for intravenous liposomal products developed with reference to an innovator liposomal product" (EMA/CHMP/806058/2009/Rev. 02) underlined. In this reflection paper, it is mentioned that "significant changes in pharmacokinetic characteristics are evident when an active substance is administered in a liposomal formulation, i.e., volume of distribution and clearance may be reduced and half-life prolonged. The clearance of the liposomal active substance is dependent on:

1. The clearance of the liposomal carrier itself
2. The rate of release of entrapped drug from the liposomal carrier
3. The clearance and metabolism of unencapsulated drug upon its release"

The changes in the formulation process of nanomedicines affects the pharmaco-kinetics (i.e., AUC, C_{max}, T_{max}) of the encapsulated bioactive molecule. In the CHMP report of the EMA regarding secreted protein acidic and rich in cysteine (SPARC) products, it is recommended that "the qualitative and quantitative composition and physicochemical properties of SPARC's proposed Doxorubicin HCl Liposome Injection and Caelyx are similar. The comparative analysis of lipid content between SPARC's product and Caelyx indicated similar content and physicochemical characteristics of lipid component. Based on the similarity of composition and physico-chemical properties, SPARC considered the proposed product as essentially similar to Caelyx" [3]. In 2006, the reflection paper published (EMEA/CHMP/79769/2006) referred that the nanosizing in nanomedicines does not correspond to novelty of the product. It seems that that novelty could be achieved by using the principles and fundamental directions of nanotechnology [3].

It is obvious that the term similarity reflects to the quality characteristics of the nanosimilar products, and quality by design (QbD) approaches need new analytical outcomes that could be complementary to those already exist. New QbD approaches are welcomed by the regulatory authorities and the scientific community should spend more time and funds to support new analytical entities in the QbD industrial process for developing nanosimilar products. It is of interest to point out that the first *off-patent* liposomal doxorubicin (doxorubicin hydrochloride liposome injection) (LipoDox) medicine has been approved by the FDA on February 4, 2013, which has been classified as the generic version of Doxil (see Table 4.2). Based on the FDA concept, generic drugs that have been approved by the FDA should meet the requirements and the quality of the prototype. The author's belief is that the similarity should be carefully taken into consideration during the approval process of liposomal medicines since the nanocarriers' behavior affects the therapeutic effectiveness and safety of the liposomal medicine. However, the *innovative excipi-ent* for liposomal nanocarriers could be an alternative approach to be discussed for approving self-assembled nanostructures such as liposomes, differentiating them from conventional generic medicines' excipients and consequently approval process.

7.3   Summary

The process of developing a therapeutic medicine requires compliance according to the guidelines to secure its safety and effectiveness.

The nanotechnological products are monitored by the FDA and EMA

Nanosimilars are the copies of the *off-patent* nanotechnological therapeutic products. The scientific discussion on *nanosimilarity* is an emerging approach that is under scientific consultation.

LipoDox is the first *off-patent* liposomal medicine that has been approved by the FDA and has been classified as the generic version of Doxil.

References

1. Ahmed I, Kaspar B, Sharma U (2012) Biosimilars: impact of biological products life cycle and European experience on the regulatory trajectory in the United States. Clin Ther 34:400–419
2. Bawa R, Johnson S (2007) The ethical dimensions of nanomedicine. Med Clin N Am 91(5):881–887
3. Demetzos C, Pippa N (2015) Fractal geometry as a new approach for proving nanosimilarity: a reflection paper. Int J Pharm 483:1–5
4. Demetzos C, Pippa N, Tountas Y (2013) Advanced therapies: new guidelines and the approval process. Pharmakeftiki 25(II):49–54
5. Directive 201/83/EC of the European Parliament and of the Council of 6 November 2001 on the Community code relating to medicinal products for human use Off J Eur Union 28 11 2001
6. Gaspar R (2007) Regulatory issues surrounding nanomedicines: setting the scene for the next generation of nanopharmaceuticals. Nanomedicine 2:143–147
7. Holloway C, Mueller-Berghaus J, Lima BS et al (2012) Scientific consideration for complex drugs in light of established and emerging regulatory guidance. Ann N Y Acad Sci 1276:26–36
8. Narang AS, Chang RK, Hussain MA (2013) Pharmaceutical development and regulatory considerations for nanoparticles and nanoparticulate drug delivery systems. J Pharm Sci 102:3867–3882
9. Regulation (EC) No 1394/2007 of the European Parliament and the Council of 13 November 2007 on advanced therapy medicinal products and amending Directive 2001/83/EC and Regulation (EC) No 726/2004. Off J Eur Union 10 12 2007
10. Resnik DB, Tinkle SS (2007) Ethics in nanomedicine. Nanomedicine 2:345–350
11. Schellekens H, Klinger E, Muhlebach S et al (2011) The therapeutic equivalence of complex drugs. Regul Toxicol Pharmacol 59:176–183
12. Stephan MT, Moon JJ, Um SH et al (2010) Therapeutic cell engineering with surface-conjugated synthetic nanoparticles. Nat Med 16:1035–1041
13. Tsiftsoglou AS, Ruiz S, Scheneider CK (2013) Development and regulation of biosimilars: current status and future challenges. BioDrugs 27(3):203–211
14. Vamvakas S, Martinaldo J, Pita R et al (2011) On the edge of new technologies (advanced therapies nanomedicines). Drug Discov Today Technol 8:e21–e28
15. van Calster G (2006) Regulating nanotechnology in the European Union. Nanotechnol Law Bus 3:359–372
16. Wagner V, Dullaart A, Bock AK, Zweck A (2006) The emerging nanomedicine landscape. Nat Biotechnol 24:1211–1217